AF847843

18
Topics in Medicinal Chemistry

Editorial Board:

P.R. Bernstein, Rose Valley, USA
A. Buschauer, Regensburg, Germany
G.I. Georg, Minneapolis, USA
J.A. Lowe, Stonington, USA
N.A. Meanwell, Wallingford, USA
A.K. Saxena, Lucknow, India
U. Stilz, Malov, Denmark
C.T. Supuran, Sesto Fiorentino, Italy

Aims and Scope

Drug research requires interdisciplinary team-work at the interface between chemistry, biology and medicine. Therefore, the new topic-related series Topics in Medicinal Chemistry will cover all relevant aspects of drug research, e.g. pathobiochemistry of diseases, identification and validation of (emerging) drug targets, structural biology, drugability of targets, drug design approaches, chemogenomics, synthetic chemistry including combinatorial methods, bioorganic chemistry, natural compounds, high-throughput screening, pharmacological in vitro and in vivo investigations, drug-receptor interactions on the molecular level, structure-activity relationships, drug absorption, distribution, metabolism, elimination, toxicology and pharmacogenomics.

In general, special volumes are edited by well known guest editors.

In references Topics in Medicinal Chemistry is abbreviated Top Med Chem and is cited as a journal.

More information about this series at http://www.springer.com/series/7355

Corey R. Hopkins
Editor

Novel Therapeutic Approaches to the Treatment of Parkinson's Disease

An Overview and Update

With contributions by

L.A. Esposito · P. Galatsis · J.L. Henderson · W.D. Hirst · B.L. Kormos · B.C. Shook

Springer

Editor
Corey R. Hopkins
Center for Neuroscience Drug Discovery
Vanderbilt University Medical Center
Nashville, Tennessee
USA

ISSN 1862-2461 ISSN 1862-247X (electronic)
Topics in Medicinal Chemistry
ISBN 978-3-319-39924-9 ISBN 978-3-319-39926-3 (eBook)
DOI 10.1007/978-3-319-39926-3

Library of Congress Control Number: 2016944160

© Springer International Publishing Switzerland 2016
This work is subject to copyright. All rights are reserved by the Publisher, whether the whole or part of the material is concerned, specifically the rights of translation, reprinting, reuse of illustrations, recitation, broadcasting, reproduction on microfilms or in any other physical way, and transmission or information storage and retrieval, electronic adaptation, computer software, or by similar or dissimilar methodology now known or hereafter developed.
The use of general descriptive names, registered names, trademarks, service marks, etc. in this publication does not imply, even in the absence of a specific statement, that such names are exempt from the relevant protective laws and regulations and therefore free for general use.
The publisher, the authors and the editors are safe to assume that the advice and information in this book are believed to be true and accurate at the date of publication. Neither the publisher nor the authors or the editors give a warranty, express or implied, with respect to the material contained herein or for any errors or omissions that may have been made.

Printed on acid-free paper

This Springer imprint is published by Springer Nature
The registered company is Springer International Publishing AG Switzerland

Preface

Parkinson's disease is the second most common neurodegenerative disease, after Alzheimer's disease, affecting ~1% of the total world's population. Despite the number of individuals afflicted with this disease, the standard of care remains L-DOPA, a drug discovered well over 40 years ago and a drug that only treats the symptoms. As the disease progresses, there is a paucity of treatments available for the patients. Over the past several years, there has been focused research in a number of areas looking to address this lack of disease-modifying treatments.

In this volume of *Topics in Medicinal Chemistry* focused on novel therapeutic approaches to the treatment of Parkinson's disease, the authors cover a range of exciting and potentially novel treatments for this devastating disease. They focus on three areas, adenosine A_{2A} antagonism, synuclein modulation, and LRRK2, and are written by leading members in their area.

The first chapter by Brian Shook (presently Enanta Pharmaceuticals, Watertown, MA) brings the reader update on the state of research in the A_{2A} antagonists for PD as well as combination A_{2A}/A_1 antagonists. The adenosine receptors are present in the indirect pathway of the basal ganglia and are thought to be beneficial in reestablishing the balance within the direct and indirect pathways. The chapter will focus on the xanthine-based and tricyclic- and bicyclic-based A_{2A} antagonists as well as dual A_{2A} and MAO inhibitors.

The second chapter by Luke Esposito (Proteotech, Kirkland, WA) explores the structural biology of α-synuclein as it pertains to PD. The readers will be caught up in the design of α-synuclein-directed therapies – those that have anti-aggregation activity as well as those that inhibit downstream toxicity of the α-synuclein aggregates.

The third chapter by Warren Hirst et al. (Pfizer Worldwide Research & Development, Cambridge, MA) provides an overview of leucine-rich repeat kinase 2 (LRRK), which is the most common known cause of autosomal dominant Parkinson's disease. Mutations in the LRRK are thought to account for ~1% of the total sporadic and ~4% of the familial cases of PD. Significant activity in the discovery and development of novel small-molecule inhibitors of LRRK by both the academic and industrial communities is discussed.

I would like to express my appreciation to the contributing authors for their work on these excellent chapters for this special volume on emerging therapeutic targets for Parkinson's disease.

Nashville, TN, USA
April 2016

Corey R. Hopkins, Ph.D.

Contents

Top Med Chem (2016) 18: 1–42
DOI: 10.1007/7355_2014_67
© Springer International Publishing Switzerland 2014
Published online: 9 August 2014

Adenosine A_{2A} Receptor Antagonists

Brian C. Shook

Abstract This chapter summarizes and updates the work on adenosine A_{2A} receptor antagonists for Parkinson's disease. Adenosine A_{2A} receptor antagonists were, and still are, a promising non-dopaminergic approach for the potential treatment of Parkinson's disease. There have been numerous publications, patent applications, and press releases that highlight new medicinal chemistry approaches to this attractive and promising target to treat Parkinson's disease. There have been many research efforts from various pharmaceutical and academic institutions targeting this receptor, and several compounds have advanced into clinical development. The chapter was broken down by scaffold type and will discuss the efforts to optimize particular scaffolds for activity, SAR, pharmacokinetics, and other drug discovery parameters. The majority of approaches focus on preparing selective A_{2A} antagonists, but a few approaches to dual A_{2A}/A_1 antagonists and A_{2A}/MAO-B will also be highlighted. The in vivo profiles of compounds will be highlighted and discussed to compare activities across different chemical series. A clinical report and update will be given on compounds that have entered clinical trials.

Keywords 6-Hydroxydopamine, A_{2A} antagonist, Adenosine, Catalepsy, Dopamine, MPTP, Parkinson's disease

Contents

B.C. Shook (✉)
Janssen Research and Development, Spring House, PA, USA
e-mail: bshook@its.jnj.com

Abbreviations

6-OHDA	6-Hydroxydopamine
APEC	(2-[(2-aminoethylamino)carbonylethyl-phenyl-ethylamino]-5′-ethylcarboxamido-adenosine)
AR	Adenosine receptors
Bu	Butyl
cAMP	Cyclic adenosine monophosphate
COMT	Catechol *O*-methyltransferase
Et	Ethyl
h	Hour(s)
i-Pr	Isopropyl
L	Liter(s)
L-DOPA	Levodopamine
MAO-B	Monoamine oxidase B
Me	Methyl
min	Minute(s)
mol	Mole(s)
MPTP	1-Methyl-4-phenyl-1,2,3,6-tetrahydropyridine
p.o.	Oral dose
PD	Parkinson's disease
Ph	Phenyl
PK	Pharmacokinetics
Pr	Propyl
rt	Room temperature
s	Second(s)
s-Bu	*sec*-Butyl
t-Bu	*tert*-Butyl

1 Introduction

Adenosine (**1**, Fig. 1) is a naturally occurring nucleoside that is present in all tissues of mammalian organisms and has numerous biological functions [1]. Early research examined how adenosine and related analogs affected heart function which resulted in the hypothesis that multiple adenosine receptors (ARs) exist. Van Calker et al. found that certain adenosine analogs were able to increase cAMP levels in

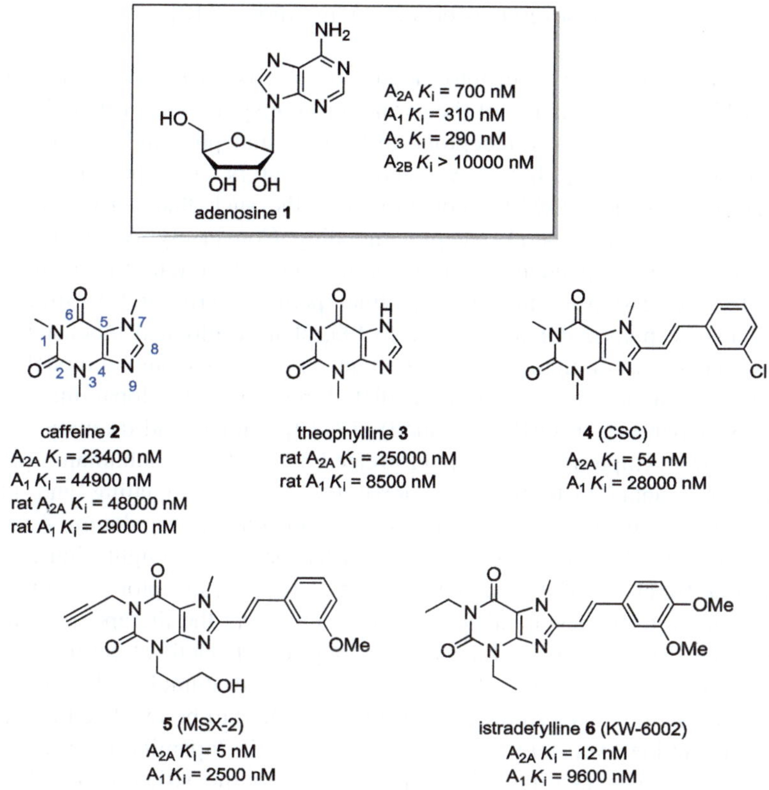

Fig. 1 Xanthine-based A_{2A} antagonists

brain tissue while other analogs inhibited cAMP accumulation, suggesting the existence of distinct adenosine receptors [2]. Adenosine is comprised of four distinct receptor subtypes designated A_1, A_{2A}, A_{2B}, and A_3 belonging to the G protein-coupled receptor superfamily [3]. Adenosine A_1 and A_3 receptors are coupled to inhibitory G proteins, while A_{2A} and A_{2B} receptors are coupled to stimulatory G proteins. The A_2 receptors were divided into two subtypes: A_{2A}, with high affinity for adenosine (0.1–1.0 μM) and A_{2B} with lower adenosine affinity (≥10 μM).

Adenosine is a neuromodulator that coordinates responses to dopamine and other neurotransmitters in areas of the brain that are responsible for motor function, mood, and learning and memory [4, 5]. Adenosine A_{2A} receptors are highly expressed in the spleen, thymus, leukocytes, and striatum but to a lesser extent in the heart, lung, blood vessels, and other areas of the brain. Although A_{2A} receptors are not brain specific, in rodent brain, autoradiography studies showed that the greatest densities of A_{2A} receptors are found in the striatum [6, 7] which closely matches the distribution in humans based on PET imaging [8, 9]. A_{2A} receptors have been cloned from several species including mouse [10], rat [11], guinea pig [12], and human [13].

The loss of dopamine input into the neostriatum is a hallmark of Parkinson's disease (PD) and causes many of the cardinal motor symptoms of this disorder. In the striatum adenosine A_{2A} receptors co-localize and physically associate with dopamine D_2 receptors [14, 15]. A_{2A} and D_2 receptors have opposing effects on adenylate cyclase and cAMP production in cells, such that activation of A_{2A} receptors inhibits dopamine D_2 receptor signaling. Conversely, A_{2A} receptor antagonists enhance D_2-dependent signaling as shown by induction of immediate early gene c-fos expression in the striatopallidal pathway [16] and facilitate other D_2-mediated responses. Of importance to PD, pharmacological blockade of A_{2A} receptors has shown dramatic beneficial effects in preclinical animal models of PD, showing potentiation of dopamine-mediated responses in dopamine depleted (6-hydroxydopamine, (6-OHDA)-treated) [17–19] animals and dramatic relief of parkinsonian symptoms in MPTP-treated [20–22] nonhuman primates [23–26]. A_{2A} antagonists facilitate dopamine receptor signaling and thereby normalize motor function in animal models of dopamine dysregulation. As a result of these findings, the adenosine A_{2A} receptor has become a sought-after target for treating PD. Blockade of A_{2A} signaling by selective A_{2A} receptor antagonists [27, 28] was shown to be beneficial for not only enhancing the therapeutic effects of L-DOPA but also reducing dyskinesia from long-term L-DOPA treatment [29–31].

The adenosine A_1 receptor is also expressed in the striatum; based on anatomical and in vivo microdialysis studies, A_1 receptors appear to be localized presynaptically of dopamine axon terminals where they inhibit dopamine release [32]. A_1 receptor antagonists facilitate dopamine release in the striatum and like A_{2A} receptors potentiate dopamine-mediated responses. Antagonism of both A_{2A} and A_1 would be synergistic – inhibition of the A_1 receptor will facilitate dopamine release, while inhibition of the A_{2A} receptor will enhance postsynaptic responses to dopamine. Interestingly, the A_1 receptor is also concentrated in neocortical and limbic system structures that are important for cognitive function. Pharmacological inhibition of A_1 receptors enhances neurotransmitter release in the hippocampus [33, 34] and enhances performance in animal models of learning and memory [35]. Antagonism of A_1 receptors could present a potential liability as it is known that some A_1 antagonists have cardiovascular [36] and diuretic [37] liabilities.

Most recently, efforts to design dual A_{2A} antagonists/MAO-B inhibitors have been reported [38, 39]. The activity and expression levels of MAO-B in the human brain increase with age [40], and the activity is associated with increased production of reactive oxygen species (ROS) which may contribute the death of neuronal cells [41]. Selegiline and rasagiline are two marketed irreversible inhibitors of MAO-B given to PD patients as monotherapy or in combination with L-DOPA [42, 43]. In addition, safinamide is a reversible inhibitor of MAO-B currently in phase III clinical trials as an adjuvant therapy for PD [44]. The proven effects of MAO-B inhibitors and the promising effects of A_{2A} antagonists in PD patients suggest that a compound having dual activities would be synergistic and could provide enhanced therapeutic and neuroprotective potential.

Reports have shown that combining positive allosteric modulators (PAMs) of metabotropic glutamate receptor 4 ($mGlu_4$) with A_{2A} antagonists produces a synergistic effect seen in the reversal of haloperidol-induced catalepsy in rats [45]. Loss of striatal dopamine signaling results in overactivation of the indirect pathway which leads to the motor impairments seen in PD [46]. Both $mGlu_4$ and A_{2A} receptors are involved in the indirect pathway and have been described as targets for pharmacological modulation thereof. As mentioned above, antagonism of A_{2A} receptors effectively reverses motor deficits in rodent models of PD, whereas the activation of $mGlu_4$, expressed presynaptically of GABAergic neurons, also reduces motor impairment in rodent models of PD [45, 47]. Although not completely understood, the observed pharmacologic interaction of these two receptors suggests that synergy may be occurring on the same and/or different neuronal pathways.

The major unmet medical needs of PD are improved symptomatic treatment without inducing adverse effects (primarily dyskinesia) associated with long-term L-DOPA or dopamine agonist therapy; opportunity to slow disease progression by protecting midbrain dopamine and other neurons from degeneration; and treatment of disease comorbidities, including cognitive dysfunction, anxiety, and depression. Based on published preclinical and human clinical data, the majority of approaches focus on selective antagonism of A_{2A} receptors, while a few approaches suggest that dual A_{2A}/A_1 receptor antagonists, dual A_{2A} antagonists/MAO-B inhibitors, and combination of A_{2A} antagonists with $mGlu_4$ PAMs may provide increased benefit to PD patients based on their synergistic and complementary pharmacology. There have been a number of excellent reviews on A_{2A} receptor antagonists reported in the literature [48–52], and this chapter will highlight key compounds, old and new, for this target including their PK and in vitro and in vivo activities. The K_i values represented in the figures and tables refer to binding affinities (human) unless noted otherwise. Clinical compounds will be noted and highlighted.

2 Xanthine-Based A_{2A} Antagonists

Adenosine (**1**) is the endogenous natural ligand for adenosine receptors and acts as an agonist with good affinity for A_{2A}, A_1, and A_3 (hA_3 $K_i = 290$ nM) receptors but is much weaker for A_{2B} (hA_{2B} $K_i > 10$ μM) (Fig. 1) [53]. Modification of the adenine and/or ribose portion of **1** resulted in numerous adenosine agonists having improved potency and metabolic stability. Unlike A_{2A} agonists, A_{2A} antagonists are devoid of the ribose moiety and generally possess a mono-, bi-, or tricyclic core structure that mimic the adenine portion of adenosine. Caffeine (**2**) and theophylline (**3**) are two naturally occurring xanthine bases, nonselective adenosine antagonists. In 1980, Fredholm et al. reported that the naturally occurring adenosine antagonists caffeine and theophylline (**2** and **3**, respectively) produced enhanced locomotor activity in mice [54].

Numerous efforts to identify xanthine-based A_{2A} antagonists have been explored to improve potency, selectivity, and aqueous solubility, a general limitation of xanthines. Chemistry efforts led to the discovery of potent A_{2A} antagonists including 3-chlorostyrylcaffeine (CSC, **4**) [55], MSX-2 (**5**) [56], and istradefylline (KW-6002, **6**) [57] having good selectivity against A_1. The selectivity against A_{2B} and A_3 is also very good for CSC (A_{2B} K_i = ND, rat A_3 $K_i > 10$ μM), MSX2 (A_{2B} K_i = ND, rat A_3 $K_i > 10$ μM), and KW-6002 (A_{2B} $K_i = 1{,}800$ nM, rat A_3 $K_i > 3{,}000$ nM). Additional xanthine-based derivatives include DMPX and CS-DMPX, structures not shown [58]. It is well established that the *trans*-styryl substituent at the 8-position is critical for A_{2A} selectivity, but also introduces a chemical liability where the olefin isomerizes from *trans* to *cis* in solution when exposed to light. It has also been shown that the olefin in one analog undergoes a [2 + 2] cycloaddition forming the corresponding dimer which was isolated and as totally inactive against the A_{2A} receptor [59]. Regardless, these compounds have served as valuable research tools, like in the case of KW-6002, a clinical compound that was recently approved in Japan for the treatment of PD [60].

Istradefylline is the most extensively studied xanthine-based A_{2A} antagonist and has shown efficacy in numerous preclinical models of PD [24, 61]. The minimum effective dose (MED) of istradefylline to reverse haloperidol-induced catalepsy in mice was 0.03 mg/kg p.o.). Istradefylline was also able to reverse parkinsonian activity in MPTP-treated primates without invoking dyskinesia. Istradefylline was also investigated in several clinical trials for PD [30, 31]. Istradefylline completed phase III clinical trials in 2009, but was not approved by the FDA due to concerns whether the efficacy finding would support the clinical use of the compound [62]. However, in March 2013, istradefylline was approved in Japan for use as an adjunctive treatment for PD, and it is marketed under the name Nouriast® [60]. The product is used to improve wearing off effects in PD patients on concomitant treatment with L-DOPA-containing medications. There is also one active phase III study using istradefylline (clinicaltrials.gov identifier: NCT01968031) and 17 completed clinical trials according to the clinicaltrials.gov website.

3 Tricyclic A_{2A} Antagonists

One of the first non-xanthine A_{2A} antagonists discovered was CGS15943 (**7**) which is very potent for A_{2A} (0.4 nM) and has high affinity for other ARs (hA_{2B} $K_i = 44$ nM, hA_3 $K_i = 95$ nM) (Fig. 2) [63]. A structurally related triazolo[4,3-*a*] quinoxaline CP-66713 (**8**) was also prepared with good affinity for A_{2A} but modest selectivity versus A_1 [64].

Schering-Plough identified a series of derivatives, where the chlorophenyl ring in **7** was replaced with a pyrazole that led to SCH-58261 (**9**) as a potent A_{2A} antagonist with modest 35-fold selectivity over the A_1 receptor (Fig. 3) [65, 66]. SCH-58261 showed efficacy in rodent models of PD after intraperitoneal (i.p.) administration, but suffers from poor solubility and is not active when dosed

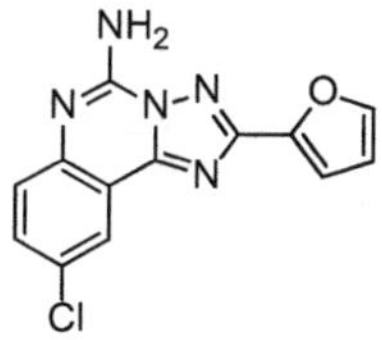

CGS 15943 (**7**)
A_{2A} K_i = 0.4 nM
A_1 K_i = 3.5 nM
rat A_{2A} K_i = 3.3 nM
rat A_1 K_i = 21 nM

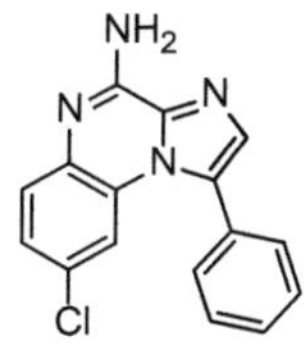

CP 66713 (**8**)
rat A_{2A} K_i = 21 nM
rat A_1 K_i = 270 nM

Fig. 2 First non-xanthine A_{2A} antagonists

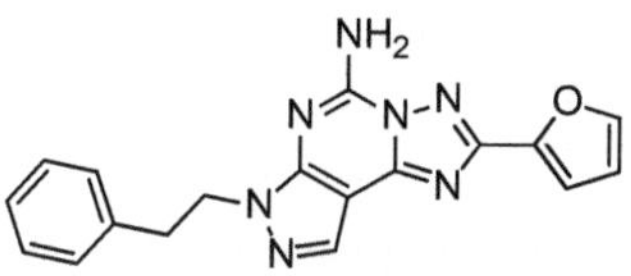

SCH-58261 (**9**)
A_{2A} K_i = 4.3 nM
A_1 K_i = 151 nM
rat A_{2A} K_i = 2.3 nM
rat A_1 K_i = 122 nM

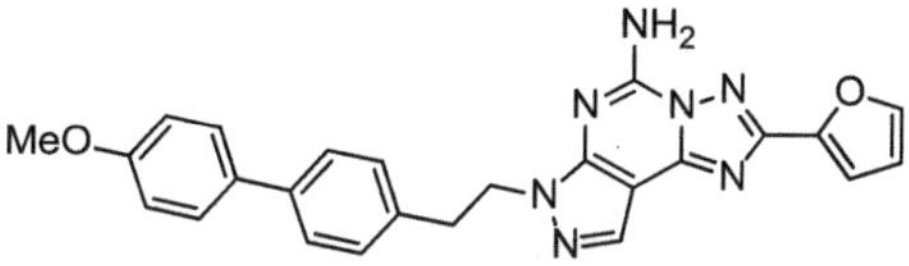

10
A_{2A} K_i = 0.5 nM
A_1 K_i = 1062 nM

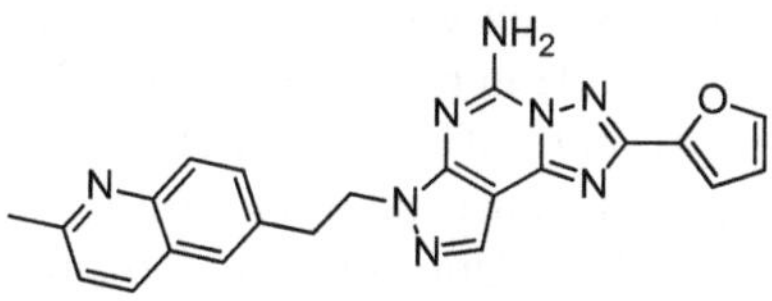

11
A_{2A} K_i = 2.4 nM
A_1 K_i = 406 nM

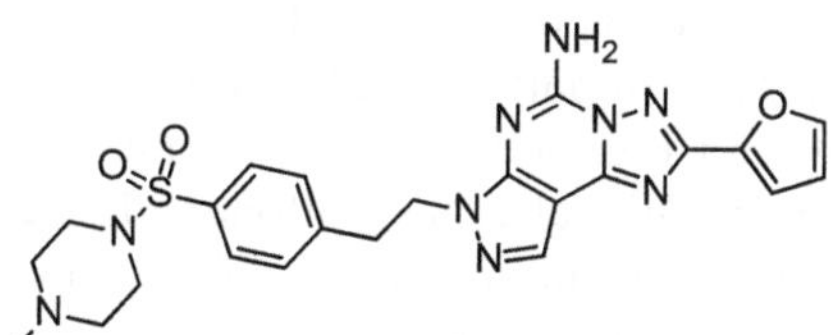

12
A_{2A} K_i = 3.8 nM
A_1 K_i = 369 nM

Fig. 3 Pyrazolo[4,3-*e*]-1,2,4-triazolo[1,5-*c*]pyrimidines

orally. It shows similar stimulatory affects to that of caffeine and exhibited positive effects in the rat 6-OHDA lesion model at 5 mg/kg i.p. SCH-58261 was widely accepted as an A_{2A} antagonist reference compound largely due to its ability to cross the blood–brain barrier (BBB). There have been numerous efforts to increase the selectivity and PK properties of this particular scaffold. A series of biaryl substituted pyrazolo[4,3-*e*]-1,2,4-triazolo[1,5-*c*]pyrimidines were also prepared [67]. The methoxy biaryl compound **10** was very potent for A_{2A} and had much better selectivity (>2,100-fold) against the A_1 receptor compared to **9**. This compound showed 30 and 5% inhibition of rat catalepsy at 1 and 4 h, respectively, at 1 mg/kg. The mediocre efficacy and short duration of action were attributed to the very low plasma exposure exhibited by this compound. Efforts to replace the methoxy aryl

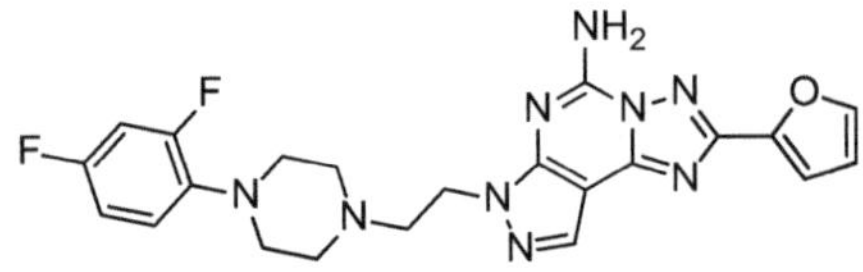

SCH-412348 (**13**)
A_{2A} K_i = 0.6 nM
A_1 K_i > 960 nM

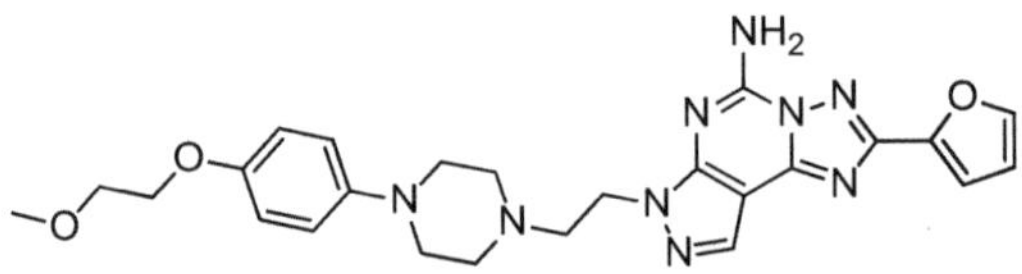

preladenant **14** (SCH-420814)
A_{2A} K_i = 1.1 nM
A_1 K_i = 1474 nM

Fig. 4 Aryl piperazine substituted pyrazolo[4,3-*e*]-1,2,4-triazolo[1,5-*c*]pyrimidines

substituent with heterocycles like pyridines and pyrazines resulted in compounds that retained potency and selectivity, but generally suffered from poor solubility and poor exposures in rat after oral dosing. Contracting the biaryl substituent into a quinoline, represented in **11**, gave potent compounds with good selectivity versus A_1. More importantly, **11** exhibited a superior PK profile in rats, compared to **9** and **10**, and reversed catalepsy in rats by 86% (1 h) and 38% (4 h) after 3 mg/kg p.o. Additionally, the sulfonamide **12** had a good balance between affinity and solubility while maintaining fairly good selectivity [68]. Unfortunately, no animal studies or PK profiles were reported for **12**.

Additional efforts to optimize potency, selectivity, and PK were achieved by replacing the phenethyl side chain (compounds **9–12**) with an aryl piperazine side chain represented in **7** (Fig. 4) [69]. A very promising compound, SCH-412348 (**13**), showed very potent A_{2A} affinity and impressive selectivity (>1,600-fold) versus A_1. Although SCH-412348 showed excellent ability to reverse haloperidol-induced catalepsy (75 and 80% inhibition at 1 and 4 h, respectively) in rats after a dose of 1 mg/kg p.o., the compound has not progressed further due to poor aqueous solubility. Further optimization led to the methoxyethoxy analog **14** (SCH-420814, preladenant), which also had excellent binding affinity for A_{2A} with good selectivity (>1,300-fold) over A_1 and improved solubility. Preladenant also displayed >1,000-fold selectivity versus A_{2B} and A_3 and was also clean in a selectivity panel of more than 60 additional targets. In rats, preladenant showed good oral exposure (AUC = 1,560 ng h/mL at 3 mg/kg p.o.) and bioavailability ($F = 57\%$) and a brain-to-plasma ratio of one. Despite the relatively short half-life ($t_{1/2} = 2.1$ h at 1 mg/kg, i.v.), preladenant (1 mg/kg) reversed haloperidol-induced catalepsy (77 and 70% inhibition at 1 and 4 h, respectively) in rats displaying good duration of action and was further characterized in multiple preclinical animal

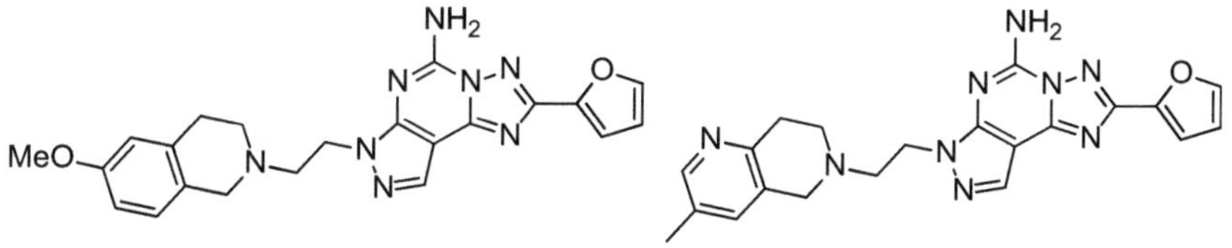

15
A$_{2A}$ K_i = 2.5 nM
A$_1$ K_i = 170 nM

16
A$_{2A}$ K_i = 2.0 nM
A$_1$ K_i = 358 nM

Fig. 5 Fused heterocyclic substituted pyrazolo[4,3-*e*]-1,2,4-triazolo[1,5-*c*]pyrimidines

models of PD [70]. Preladenant showed significant activity when dosed orally in the rat hypolocomotion model (minimum effective dose 0.1 mg/kg p.o.), 6-OHDA-lesioned rat model (minimum effective dose 0.03 mg/kg p.o.), and MPTP-treated monkey model [50]. Preladenant showed efficacy in PD patients in one phase II clinical trial (P4501) at 2, 5, and 10 mg/kg p.o. Unfortunately on May 13, 2013, Merck (Schering-Plough) announced that the development of preladenant was discontinued due to lack of efficacy in three phase III clinical trials for PD [71].

A variety of fused heterocyclic derivatives of the triazolopyrimidine scaffold were prepared that had comparable A$_{2A}$ binding affinities with respect to the aryl piperazines **13** and **14**, but were generally much less selective against the A$_1$ receptor (Fig. 5) [72]. The tetrahydroisoquinoline derivative **15** was 68-fold selective versus A$_1$, but had excellent binding affinity for A$_{2A}$. The compound had moderate plasma exposure in rats and maintained the ability to reverse catalepsy over time by 50 and 58% at 1 and 4 h time points, respectively, after a 3 mg/kg dose p.o. A related azatetrahydroisoquinoline **16** had higher plasma exposure (>fivefold) in rats compared to **15**, good bioavailability ($F = 71\%$), low clearance (4.7 mL/min/kg), and a half-life of 11.3 h. Compound **16** reversed catalepsy in rat (3 mg/kg) by 85 and 30% at 1 and 4 h, respectively, but was less effective at the 4 h time point compared to preladenant despite having a longer half-life. This could be explained by lower brain exposure/free fraction, but brain levels were not reported.

There were significant efforts to identify a suitable replacement for the commonly featured furan, which has the potential to undergo oxidative metabolism. This was a challenging effort and the majority of aryl and heteroaryl replacements retained A$_{2A}$ potency but had significant reduction in the selectivity versus A$_1$. However, the benzoxazole derivatives **17** and **18**, containing the pyrazine and alkyne replacements, respectively, maintained good potency and excellent selectivity over A$_1$ (Fig. 6) [73–75]. Compound **17** was claimed in a patent application, but only limited biological data was reported. The alkyne derivative **18** reversed catalepsy in rats by 55 and 48% at 1 and 4 h, respectively, after a 1 mg/kg dose. This represents one of the few compounds from this series that does not have a furan, but is able to maintain good potency in vitro and in vivo, excellent selectivity, and good aqueous solubility [76].

Attempts to further optimize the preladenant structure replaced the pyrazole ring with an imidazole ring to give a series of 3*H*-[1,2,4]-triazolo[5,1-*i*]purin-5-amines

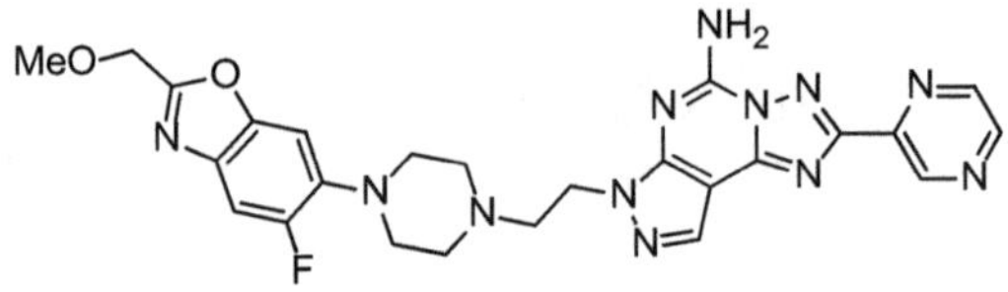

17
A_{2A} K_i = 1.1 nM
A_1 K_i = 1900 nM

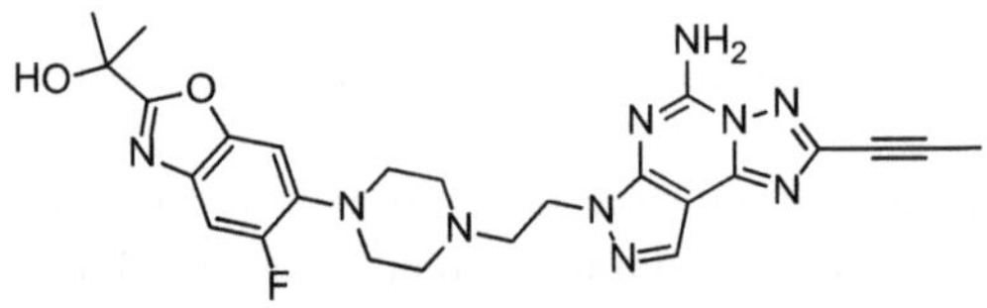

18
A_{2A} K_i = 0.4 nM
A_1 K_i = 1722 nM

Fig. 6 Furan replacements of aryl piperazine substituted 3*H*-[1,2,4]-triazolo[5,1-*i*]purin-5-amines

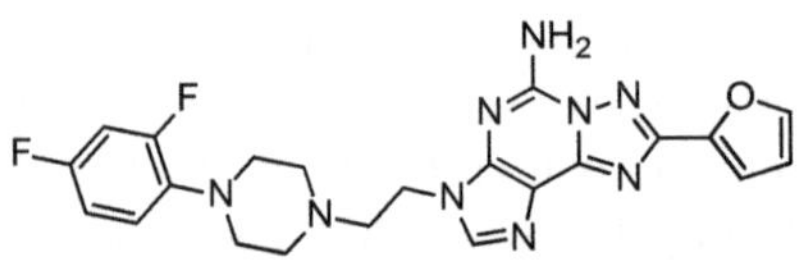

19
A_{2A} K_i = 0.7 nM
A_1 K_i = 62 nM

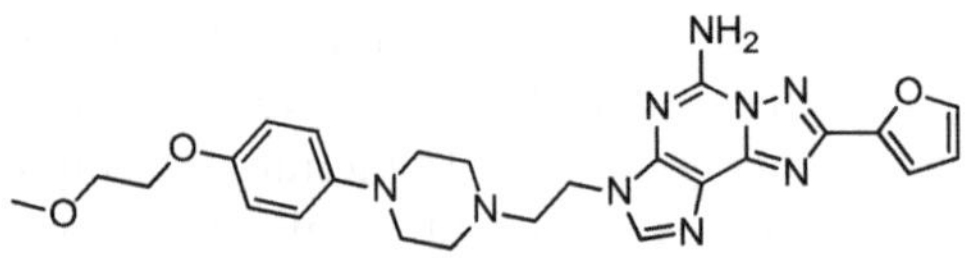

20
A_{2A} K_i = 0.9 nM
A_1 K_i = 602 nM

Fig. 7 Aryl piperazine substituted 3*H*-[1,2,4]-triazolo[5,1-*i*]purin-5-amines

(Fig. 7) [77]. Compounds **19** and **20** contain the optimal aryl piperazine substituents present in the pyrazolopyrimidines **13** and **14**. Both **19** and **20** have good affinities for A_{2A} while maintaining good selectivity versus A_1 (>80- and >660-fold, respectively), albeit considerably lower than the selectivities obtained from **13** and **14**. Compound **20** reversed catalepsy in rats by 55 and 50% at 1 and 4 h, respectively, after a 1 mg/kg oral dose. However, the compounds from this series were inferior to the corresponding compounds from the pyrazolopyrimidine scaffold. Efforts to replace the furan substituent with substituted aryls and heteroaryls resulted in compounds having good A_{2A} potency but were much less selective versus A_1 receptors.

A variety of ethylamino derivatives were prepared in a pyrazolo[4,3-*e*]-1,2,4-trizolo[4,3-*c*]pyrimidon-3-one series (Fig. 8) [78]. The ethylene dimethylamino analog **21** has good binding affinity for A_{2A} and is 269-fold selective versus A_1 receptors. In general, analogs had good binding affinities but were not able to reverse haloperidol-induced catalepsy in rat at 10 mg/kg p.o. Tissue distribution

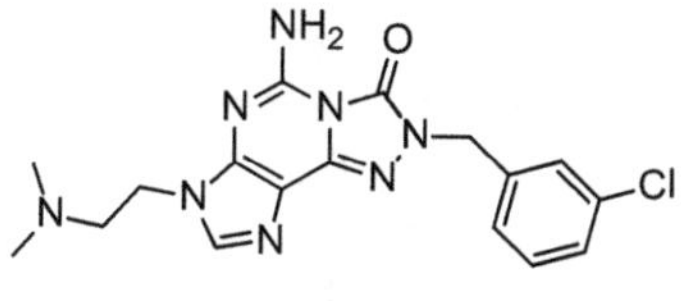

21
A_{2A} K_i = 5.2 nM
A_1 K_i = 1399 nM

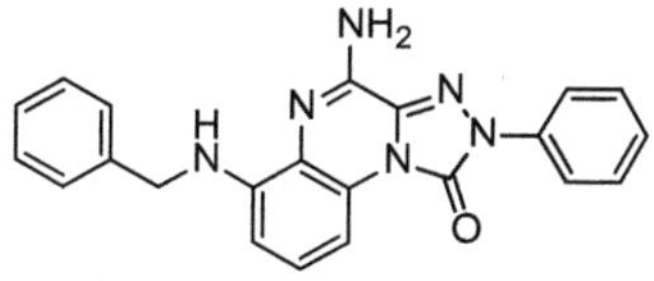

22
bovine A_{2A} K_i = 6.5 nM
bovine A_1 K_i = 730 nM

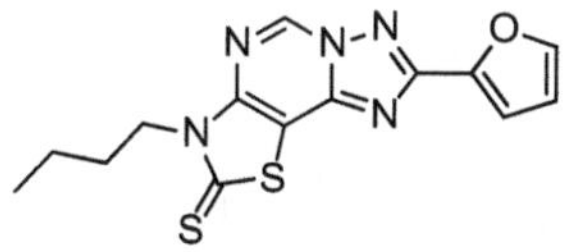

23
A_{2A} K_i = 0.1 nM
A_1 K_i = 148 nM

Fig. 8 Pyrazolo[4,3-*e*]-1,2,4-trizolo[4,3-*c*]pyrimidon-3-one, triazoloquinoxalinones, and thiazolotriazolopyrimidines

studies showed very low exposures in rat brain. A series of triazoloquinoxalinones were prepared to be selective A_{2A} antagonists with **22** being the most potent compound reported [79, 80]. Substitution on the benzylic position generally decreased the activity as did replacement of the phenyl with heteroaryl rings. No PK or in vivo data was reported for this series of compounds. A series of thiazolo-triazolopyrimidines was synthesized, and multiple compounds showed subnano-molar binding affinities [81]. Compound **23** had potent A_{2A} binding affinity and exhibited good selectivity against A_1 receptors. Compound **23** also showed very potent functional (cAMP) activity ($K_i = 0.1$ nM) and was able to reverse both haloperidol-induced catalepsy and akinesia in mice after a dose of 10 mg/kg p.o.

A series of methylene amine substituted arylindenopyrimidines were reported as dual A_{2A}/A_1 receptor antagonists (Fig. 9) [82]. Compound **24** was the original lead compound that was potent in both A_{2A} and A_1 functional assays (no binding affinity has been reported) and had an ED_{50} of 5.0 mg/kg p.o. in the haloperidol-induced catalepsy model in mice. Compound **24** suffered from poor solubility and it was found to be Ames positive, presumably due to metabolic liability of the furan. A variety of heterocyclic furan replacements were prepared that generally maintained good in vitro potency for both A_{2A} and A_1, but were not active in the mouse catalepsy model. Replacing the furan with phenyl gave **25** that had good in vitro potency and comparable in vivo activity ($ED_{50} = 8.0$ mg/kg p.o.) reversing catalepsy in mice. Compound **25** was negative in the Ames screen which somewhat confirmed that the furan was responsible for the positive Ames, but still suffered from poor solubility. The phenyl ring seemed to be a suitable replacement for the furan and was devoid of the Ames liability, so a variety of amines were

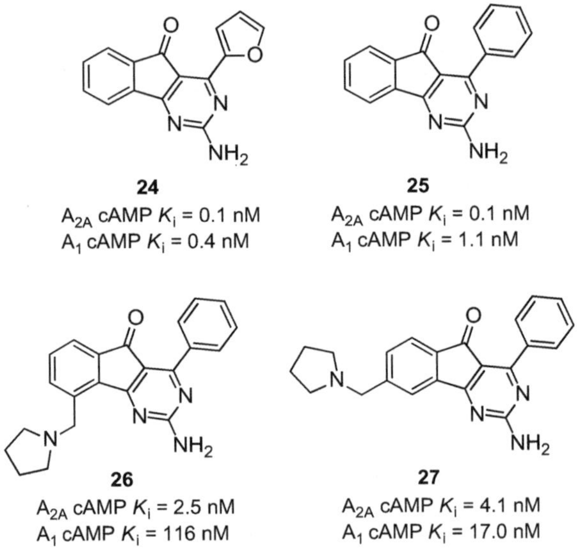

Fig. 9 Arylindenopyrimidines

incorporated at the 9-position of the scaffold to increase solubility. These efforts resulted in the synthesis of **26** that had good in vitro potency and had an ED_{50} of 3.8 mg/kg p.o. in the mouse catalepsy model. Moving the pyrrolidine to the 8-position gave compound **27** that was equipotent at A_{2A}, more potent at A_1, and significantly increased potency in vivo ($ED_{50} = 0.2$ mg/kg p.o.) in the mouse catalepsy model. Compound **27** had good bioavailability and exposure in mice, rats, and monkeys and had a brain-to-plasma ratio of 3 in rats. Further characterization of **27** showed that it was active in rat catalepsy ($ED_{50} = 0.5$ mg/kg p.o.) and had minimum effective doses of 1.0, 1.0, and 10 mg/kg for reserpine-induced akinesia model in mice, 6-OHDA lesion model in rats, and reversing motor disability in MPTP-treated marmosets, respectively [83].

Further evaluation revealed that metabolism of **27** resulted in the formation of reactive metabolites that were attributed to some adverse events seen in the 28-day GLP toxicology studies in nonhuman primates [84]. Metabolite identification studies showed that two reactive metabolites were formed, an endocyclic iminium ion in the pyrrolidine ring and an arylaldehyde formed by oxidation at the benzylic position [85]. The development of **27** was discontinued based on these findings, and the focus was to identify a suitable backup compound devoid of the metabolic liabilities [86]. Compound **28** is a representative amide that had good functional activity at both A_{2A} and A_1 receptors and reversed haloperidol-induced catalepsy in mice ($ED_{50} = 0.4$ mg/kg p.o.) (Fig. 10). PK studies showed that **28** had good exposures in mice, rats, and monkeys and had a brain-to-plasma ratio of ~0.5. Further characterization showed that **28** effectively reversed catalepsy in rats ($ED_{50} = 0.6$ mg/kg p.o.), had efficacy in mouse and rat catalepsy models for at

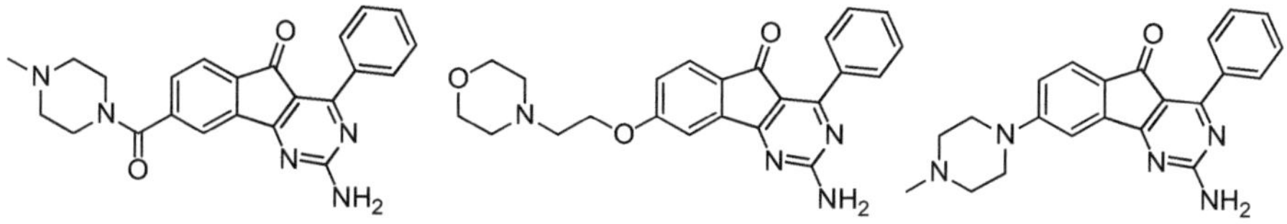

28
A_{2A} cAMP K_i = 8.2 nM
A_1 cAMP K_i = 58.4 nM

29
A_{2A} cAMP K_i = 6.5 nM
A_1 cAMP K_i = 48.2 nM

30
A_{2A} cAMP K_i = 2.7 nM
A_1 cAMP K_i = 35.6 nM

Fig. 10 Substituted arylindenopyrimidines

least 4 h after a 10 mg/kg dose, and had minimum effective doses of 1 mg/kg p.o. in the mouse and rat reserpine models and in the 6-OHDA lesion model in rats. The ether-linked compound **29** was very potent in vivo with an $ED_{50} < 0.1$ mg/kg p.o. in the mouse catalepsy model. A tissue distribution study in rats (10 mg/kg p.o.) showed that **29** had a brain C_{max} of 4.1 μM and a brain-to-plasma ratio of 3. Good exposures were also reported in mice and monkeys. Like **28**, compound **29** exhibited excellent in vivo activity (minimum effective dose in each model was 1 mg/kg p.o.) in a variety of PD models including rat catalepsy ($ED_{50} = 0.3$ mg/kg p.o.), mouse and rat reserpine models, and the 6-OHDA model in rats. The amino-linked compound **30** was potent in vitro and reversed catalepsy in mice with an $ED_{50} < 1.0$ mg/kg p.o. All of the compounds were devoid of the metabolic liabilities associated with compound **27**.

4 Bicyclic A_{2A} Antagonists

ZM241385 (**31**) is a potent A_{2A} antagonist that has good selectivity versus A_1 and has been widely used as a research tool (Fig. 11) [87]. This compound has favorable aqueous solubility but is not orally active in rodent models of PD. In 2008, a co-crystal structure of ZM241385 bound to the A_{2A} receptor was reported [88]. The structure revealed that the commonly featured furan moiety sits deep in the ligand-binding cavity where the oxygen forms a hydrogen bond to an asparagine residue. Compound **32** is a selective A_{2A} antagonist that reverses catalepsy in mice at 3 mg/kg p.o [89]. Interestingly, the des-methyl version has a similar in vitro profile, but is significantly less active in the mouse catalepsy model. One additional methylene between the NMe and piperazine resulted in reduced potency at A_{2A}. Only fluoro substituted phenyls were reported, so it is unclear how other substitution might affect activity and selectivity.

Substituted 1,2,4-triazolo[1,5-*c*]pyrimidines are a bicyclic system related to preladenant without the fused pyrazole ring (Fig. 12) [90]. A variety of linkers were explored including O, S, NH, NMe, and NEt corresponding to compounds **33–37** while keeping the optimized aryl piperazine, present in preladenant, fixed. In general, all of the linkers maintained good affinity for A_{2A} and exhibited good

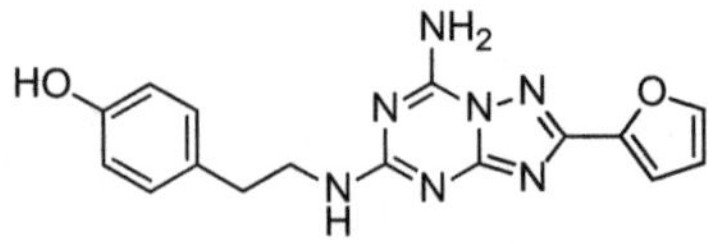

ZM-241385 (**31**)
A_{2A} K_i = 0.8 nM
A_1 K_i = 255 nM

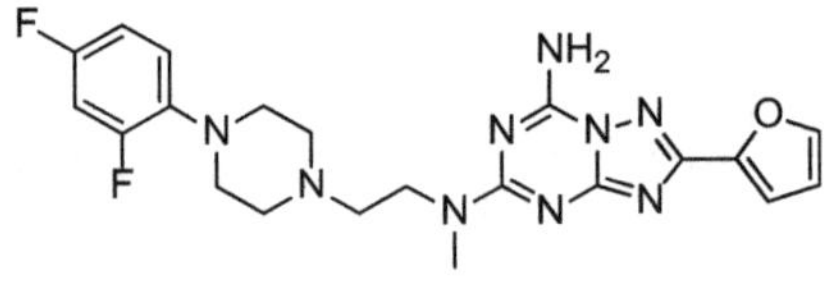

32
rat A_{2A} K_i = 4 nM
rat A_1 K_i = 820 nM

Fig. 11 Triazolotriazines

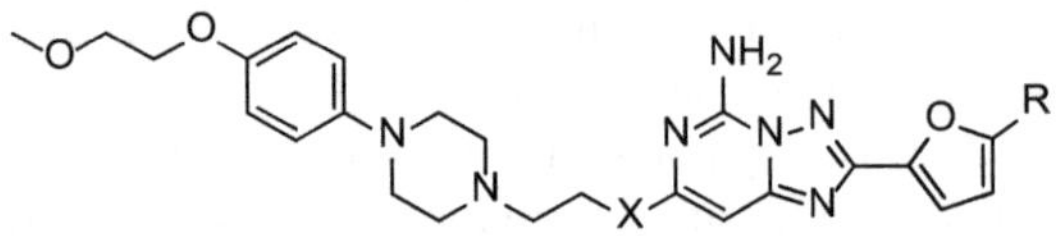

Compound	R	X	A_{2A} K_i (nM)	A_1 K_i (nM)	rat plasma AUC at 3 mg/kg, po, ng h/mL
33	H	O	2.8	1134	943
34	H	S	1.5	1448	965
35	H	NH	9.5	1376	103
36	H	NMe	2.5	1735	521
37	H	NEt	1.5	153	230
38	Cl	O	40	1600	6

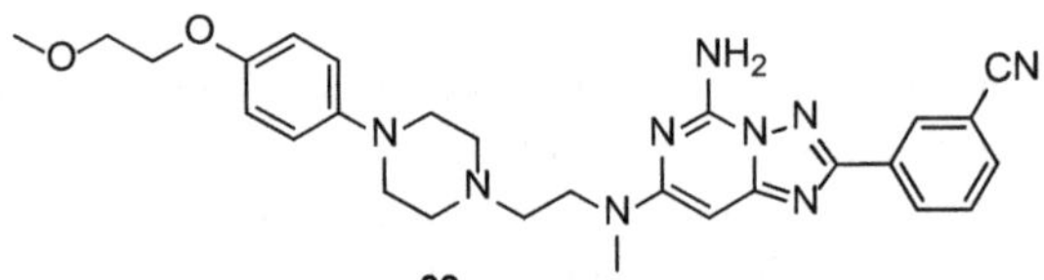

39
A_{2A} K_i = 0.7 nM
A_1 K_i = 750 nM
Rat AUC (3 mg/kg) = 546 ng h/mL

Fig. 12 Substituted 1,2,4-triazolo[1,5-*c*]pyrimidines

selectivity versus A_1, but the O and S linkages gave the best plasma exposures in rats compared to any of the nitrogen-linked analogs. Interestingly, a chlorine substituent (**38**) on the furan significantly reduced A_{2A} activity and dramatically reduced plasma exposure compared to the *des*-chloro analog **33**. Replacement of the furan in **33** with 3-cyanophenyl afforded compound **39** that had excellent binding affinity for A_{2A}, had good selectivity versus A_1, and had nearly identical exposure in rats. The compounds from this series were evaluated in the haloperidol-

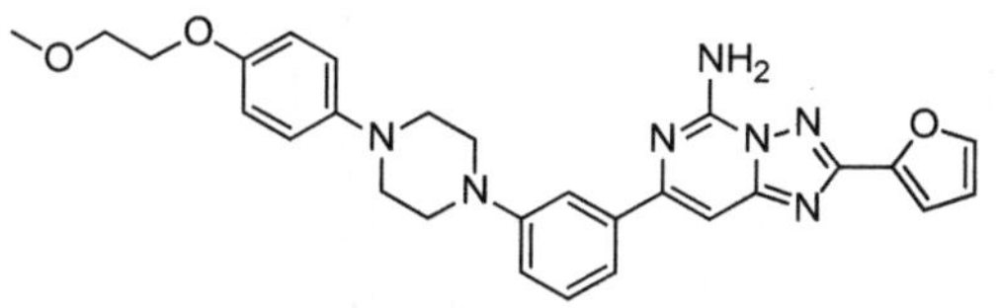

40
A_{2A} K_i = 1.0 nM
A_1 K_i = 1059 nM

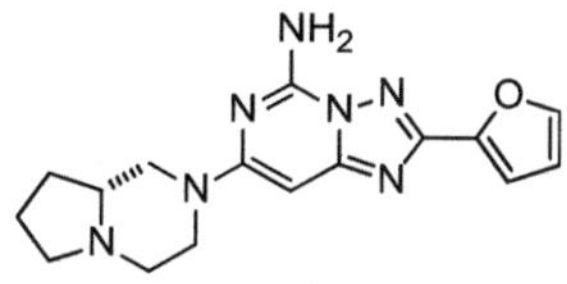

41
rat A_{2A} K_i = 63 nM
rat A_1 K_i = 1071 nM

Fig. 13 Triazolopyrimidines

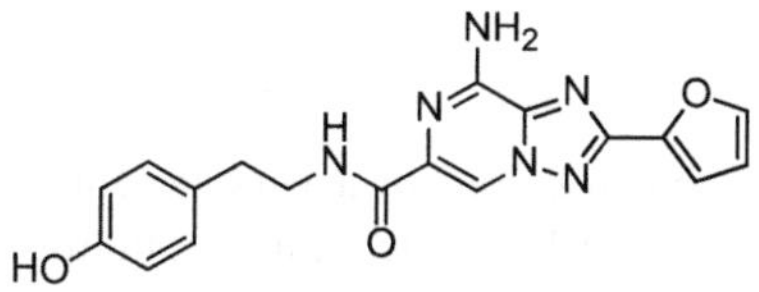

42
rat A_{2A} K_i = 1.0 nM
rat A_1 K_i = 320 nM

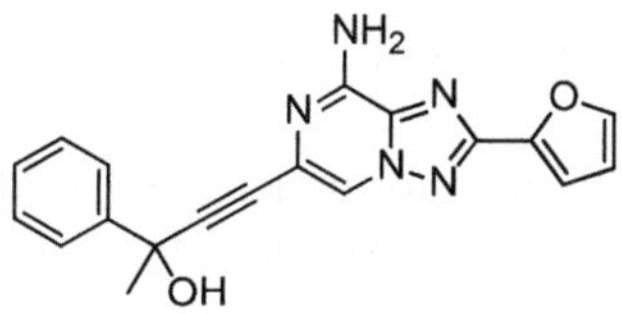

43
rat A_{2A} K_i = 12 nM
rat A_1 K_i = 41 nM

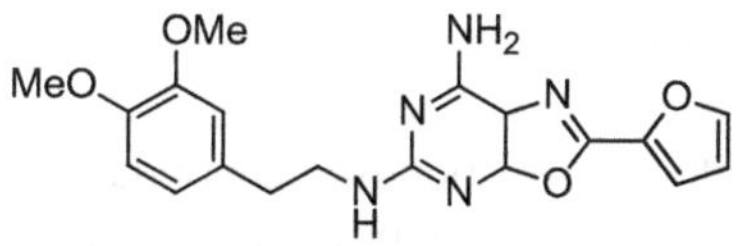

44
porcine A_{2A} K_i = 14 nM
porcine A_1 K_i = 1270 nM

Fig. 14 Triazolopyrazines and oxazolopyrimidines

induced catalepsy model in rats. Compound **34** gave the best results with 65% inhibition of catalepsy at both 1 and 4 h after 3 mg/kg p.o.

A related series of substituted 7-phenyl-[1,2,4]-triazolo[1,5-*c*]pyrimidine-5-amines were also reported (Fig. 13) [91]. Compound **40** was identified as having potent affinity for A_{2A} and good selectivity versus A_1. Further evaluation showed that **40** significantly reversed haloperidol-induced catalepsy at 3 mg/kg p.o. (55 and 40% inhibition at 1 and 4 h, respectively). It also reversed CGS-21680-induced hypolocomotion in rats by 59% after a 1 mg/kg p.o. The much smaller bicyclic substituted compound **41** and related analogs generally had reduced in vitro potency compared to larger substituents like those found in **40** [92]. However, **41** was able to reverse catalepsy in rats for at least 2 h after a 3 mg/kg oral dose. The *R*-stereochemistry of bicycle proved to be significant as it exhibited greater in vitro potency and in vivo efficacy than the corresponding *S*-enantiomer.

A series of related triazolopyrazines have been reported as potent A_{2A} antagonists having good selectivity versus A_1 (Fig. 14). Compound **42** incorporates the

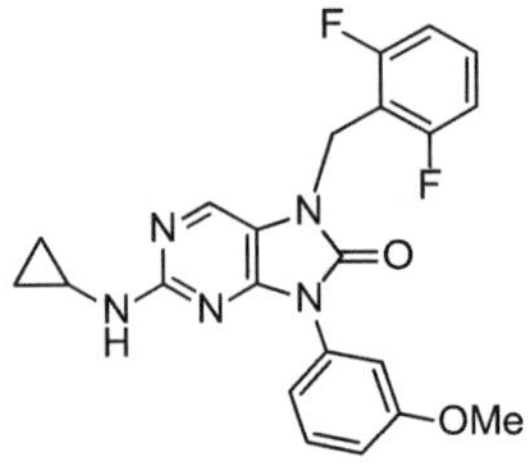

45
A_{2A} K_i = 4 nM
A_1 K_i = 236 nM

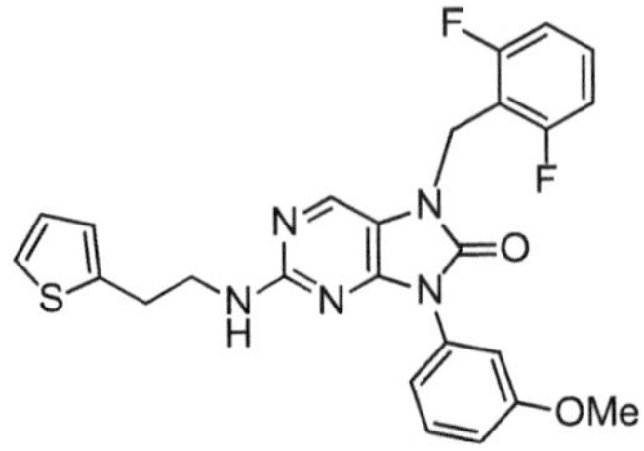

46
A_{2A} K_i = 9 nM
A_1 K_i = 1967 nM

Fig. 15 Substituted purinones

amino-ethylphenol unit from ZM241385 that resulted in a very potent A_{2A} antagonist having >300-fold selectivity against A_1 [93]. Side chains containing cyclic amides like morpholine and piperazine resulted in steep decreases in A_{2A} activity. No in vivo characterization of this series of compounds was reported. A distinctly different series of alkyne substituted triazolopyrazines were reported and are generally represented by compound **43** [94]. Typically, compounds having this type of substitution exhibited good affinity for A_{2A} but were very modestly selective versus A_1. Interestingly the corresponding *des*-methyl analog of **43** had ~30-fold decrease in affinity for A_{2A}, and this trend was consistent for all tertiary alcohols compared to their corresponding secondary analogs. Compound **43** was effective in reversing catalepsy in both mice and rats, having an $ED_{50} = 3$ mg/kg p.o. for each. This compound also showed good activity in the 6-OHDA model in rat having a minimum effective dose of 3 mg/kg p.o. A related series of oxazolopyrimidines were prepared in an effort to find a suitable A_{2A} antagonist that could be used as a PET tracer [95]. Compound **44** was the most potent and selective A_{2A} antagonist reported from this series, but this series of compounds was not further developed as PET tracers due to their high degree of unspecific binding which was determined by autoradiography experiments.

A series of trisubstituted purinones was discovered and investigated from a high-throughput screening (Fig. 15) [96]. Substituting N-7 and N-9 with various aryl and benzyl substituents while keeping the cyclopropyl amine substitution at the 2-position constant exhibited some interesting SAR depending on the substitution of the aryl rings. Compound **45** was the most potent compound for A_{2A} and was >50-fold selective versus A_1 receptors. Replacement of the cyclopropyl amine with 2-(thiophen-2-yl)ethanamine afforded compound **46** that had comparable activity for A_{2A} but was considerably more selective (>200-fold) versus A_1 compared to compound **45**. The thiophene analog was the most potent compound reported and had good functional activity (cAMP) in rat ($K_i = 18$ nM). No PK or in vivo results were reported for any compounds within this series.

The (-)-(*R*,*S*)-enantiomer of antimalarial drug mefloquine (**47**) was found to have modest A_{2A} activity while researchers were investigating the compound's

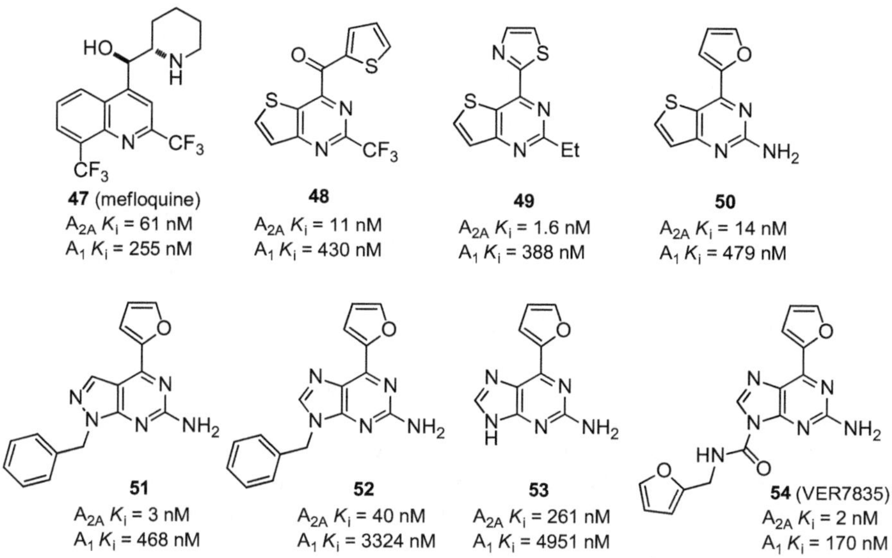

Fig. 16 Thieno[3,2-*d*]pyrimidines, pyrazolo[3,4-*d*]pyrimidines, and 6-arylpurines

side effects (Fig. 16) [97]. Compound **47** was not selective against A_1 and had no in vivo activity in rodent models of PD, but it did serve as an initial lead to develop novel A_{2A} antagonists. Optimization of this lead by changing the central scaffold resulted in the synthesis of a series of thieno[3,2-*d*]pyrimidines having good affinity for A_{2A} [98, 99]. Compound **48** represents one of the most potent compounds from this ketoaryl series and has modest selectivity versus A_1. In general, aryl and pyridyl keto substituents were significantly less potent compared to electron-rich 5-membered heterocycles. Direct attachment of the 2-thiazolyl ring in **49** resulted in significant increases in A_{2A} binding affinity ($K_i = 1.6$ nM) and selectivity over A_1. Compound **49** was also able to reverse hypolocomotion in mice at 30 mg/kg i.p. A similar aminopyrimidine **50** was identified that had good receptor binding for A_{2A} but was moderately selective versus A_1. Efforts to identify other bicyclic templates identified pyrazolo[3,4-*d*]pyrimidine **51** and 6-arylpurines **52** and **53** [100]. Compounds **51** and **52** both show good affinity for A_{2A} and were both able to reverse haloperidol-induced hypolocomotion in mice at 30 mg/kg i.p., but they were not active when dosed orally. Interestingly, the 6-arylpurine **53** had much weaker affinity for A_{2A} but was effective in reversing hypolocomotion in mice when dosed at 10 mg/kg i.p. and 1 mg/kg p.o. No plasma or brain exposures were reported to explain this unexpected result. Further optimization led to compound **54** (VER7835) which was very potent for A_{2A} and maintained modest selectivity versus A_1. Compound **54** was able to reverse haloperidol-induced hypolocomotion in mice after a 10 mg/kg oral dose, but low oral bioavailability and poor in vivo stability halted any further development of this series of compounds.

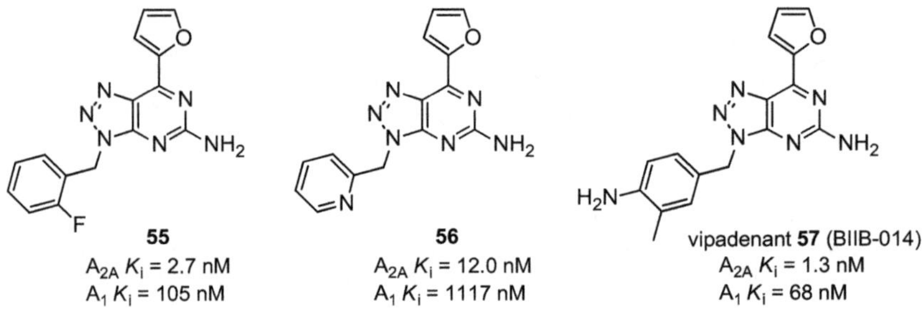

Fig. 17 Benzyl substituted triazolo[4,5-*d*]pyrimidines

The previous work led to an optimization effort to find more potent orally active compounds with good PK profiles (Fig. 17). A variety of substituted benzyl analogs were prepared that had good A_{2A} activity but were not able to reverse haloperidol-induced hypolocomotion when dosed orally [101]. Compound **55** had potent A_{2A} activity and exhibited moderate selectivity versus A_1, but its inactivity was attributed to low solubility and poor PK profiles and led to the synthesis of heterocyclic compounds like the 2-pyridyl analog **56**. Despite good in vitro activity and increased solubility, **56** was not active in vivo if given 30 mg/kg p.o. Further optimization of this scaffold led to the discovery of vipadenant (**57**) (V2006/BIIB014) which was evaluated in phase II clinical trials [102]. Vipadenant showed excellent binding affinity for A_{2A} and had modest selectivity against A_1 ($K_i = 68$ nM) and A_{2B} ($K_i = 63$ nM), while it showed good selectivity against A_3 ($K_i = 1{,}005$ nM). When tested in vivo in the mouse and rat haloperidol-induced hypolocomotion models, **57** had a minimum effective dose of 0.1 and 1 mg/kg, respectively. Vipadenant was able to increase contralateral rotations in 6-OHDA-lesioned rats when dosed orally in combination with apomorphine at 3 and 10 mg/kg [48]. The compound also showed reversal of motor disability in MPTP-treated marmosets without dyskinesias at a minimum effective dose of <5 mg/kg p.o [48]. Vipadenant also exhibited a favorable brain-to-plasma ratio as indicated by a 57% brain uptake in rats after a 2 mg/kg i.v. dose. The excellent in vivo data and a favorable safety profile helped progress this compound through phase II clinical trials. Despite positive phase II clinical results, in July 2010 Vernalis and Biogen have discontinued the development of vipadenant due to some preclinical toxicology findings that showed adverse side effects [103]. Vernalis is developing V81444 (structure not disclosed) which recently completed a phase I study in healthy male volunteers (clinicaltrials.gov identifier: NCT01634568). In December 2012, they reported positive results from a receptor occupancy study with this compound and that a phase IIa clinical trial was planned to begin by the second quarter of 2013 [104]. There are no reports indicating that this study is ongoing or has even started.

A series of triazolo-9*H*-purines were reported as potent A_{2A} antagonists having modest to low selectivity versus A_1 (Fig. 18) [105]. Compounds **58** and **59** (ST-1535) are representative examples of the series developed by Sigma-Tau.

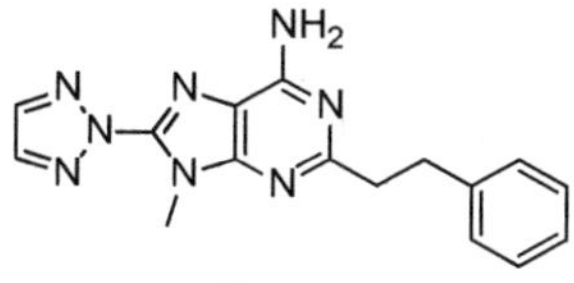

58
A_{2A} K_i = 4.7 nM
A_1 K_i = 80.0 nM

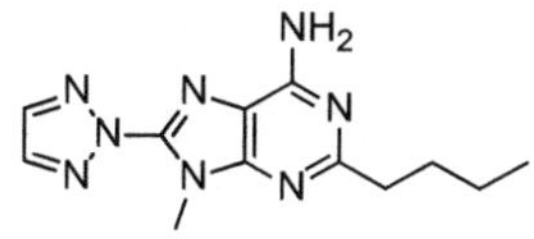

59 (ST-1535)
A_{2A} K_i = 6.6 nM
A_1 K_i = 71.8 nM

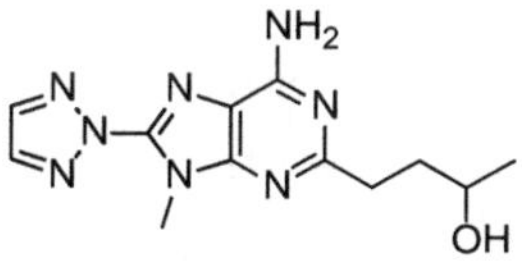

60 (ST-3932)
A_{2A} K_i = 8 nM
A_1 K_i = 24 nM

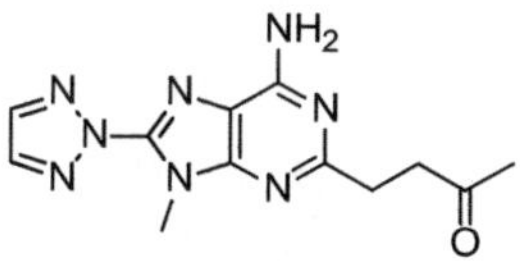

61 (ST-4206)
A_{2A} K_i = 12 nM
A_1 K_i = 192 nM

Fig. 18 Triazolo-9*H*-purines

ST-1535 has been characterized extensively in various animal models of PD. ST-1535 has minimum effective doses of 5.0 and 1.25 mg/kg in hypolocomotion and haloperidol-induced catalepsy models in mice, respectively [106]. Further studies show that ST-1535 potentiated L-DOPA activity in 6-OHDA-lesioned rats [107, 108]. In the MPTP-treated marmosets, ST-1535 dosed alone at 40 mg/kg significantly increased locomotor activity but did not improve motor disability [109]. However, when dosed at 20 mg/kg in combination with L-DOPA (2.5 mg/kg), a significant reversal of motor disability was seen compared to 20 mg/kg ST-1535 alone. The significance was not reported with respect to L-DOPA (2.5 mg/kg) alone. In addition, ST-1535 reduced tacrine-induced tremulous jaw movements in rats and was also studied for neuroprotective potential in rats. ST-1535 has been studied in phase I clinical trials, but it is no longer listed on Sigma-Tau's pipeline. A recent patent application revealed that compounds **60** and **61** are active metabolites of **59** and had similar A_{2A}, but had low selectivity versus A_1 [110]. Both metabolites were effective in reversing catalepsy in mice at 10 mg/kg p.o. and increased contralateral rotations in 6-OHDA-treated rats at 3 mg/kg p.o. It is not exactly clear why the development of these compounds was halted.

A series of aminomethyl substituted thieno[2,3-*d*]pyrimidines has been reported as A_{2A} antagonists having moderate selectivity versus A_1 (Fig. 19) [111]. The in vitro activities for A_{2A} and A_1 are reported as functional activity, not binding affinity. The unsubstituted furan proved to be the optimal substituent for A_{2A} potency and selectivity versus A_1. The basicity of amine substituent was the key to obtain in vivo activity in the haloperidol-induced catalepsy model in mice. Very basic amines like pyrrolidines and piperidines had good in vitro potency but were not able to reverse catalepsy in vivo at 10 mg/kg p.o. The morpholine compound **62** was much less basic (calcd $pK_a = 6.5$) than the corresponding pyrrolidines (calcd $pK_a = 8.2$) and piperidines (calcd $pK_a = 8.5$) and had a reported oral ED_{50} of

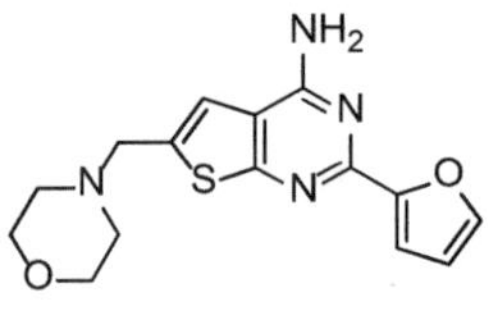

62

A_{2A} cAMP K_i = 29.0 nM
A_1 cAMP K_i = 1680 nM

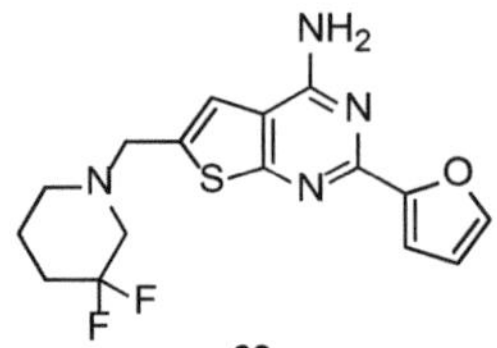

63

A_{2A} cAMP K_i = 6.6 nM
A_1 cAMP K_i = 290 nM

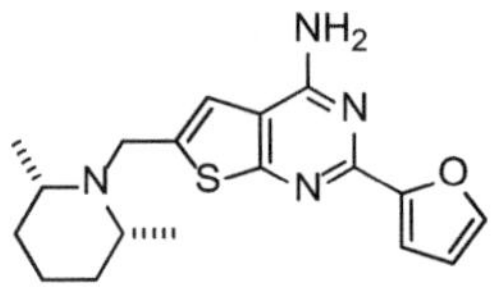

64

A_{2A} cAMP K_i = 5.3 nM
A_1 cAMP K_i = 100 nM

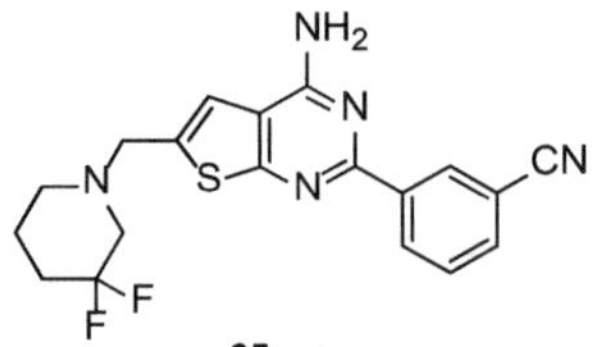

65

A_{2A} cAMP K_i = 39.0 nM
A_1 cAMP K_i = 840 nM

Fig. 19 Aminomethyl substituted thieno[2,3-*d*]pyrimidines

1.3 mg/kg in the mouse catalepsy model. A variety of less basic amines were explored to determine the scope and to identify if the basicity was indeed a real trend. Compounds **63** and **64** were even more potent in vitro than **62** and significantly reversed catalepsy at 10 mg/kg p.o. Further characterization of **63** showed that it had an ED_{50} of <1 mg/kg p.o. in mouse catalepsy, but suffered from a suboptimal PK profile that was attributed to metabolism of the furan. Efforts to substitute the 5-position of the furan with methyl, ethyl, isopropyl, cyclopropyl, $-CHF_2$, chlorine, and bromine generally resulted in slight decreases of in vitro potency, but dramatic decreases of in vivo activity. Replacing the furan with various 5-membered heterocycles resulted in a very large decrease of in vitro potency. Interestingly, the furan could be replaced with a 3-cyanophenyl group and maintain good A_{2A} activity. Compound **65** was ~6-fold less potent than the corresponding furan analog, but maintained good activity in vivo (ED_{50} < 1 mg/kg) and had a superior PK profile. Various donating and withdrawing substituents on the aryl ring were explored but only the 3-CN substituted compounds were able to maintain desirable A_{2A} activity.

A similar series of benzyl substituted thieno[2,3-*d*]pyrimidines has also been shown to have good A_{2A} activity and have varying degrees of selectivity versus A_1 (Fig. 20) [112]. Compound **66** was the representative example for this series having good A_{2A} activity and an ED_{50} of 2.2 mg/kg p.o. in the mouse catalepsy model. This compound had good oral exposures in rodent, but suffered from high clearance, a short half-life, and a short duration of action in mouse catalepsy. Compound **66** did, however, have a brain-to-plasma ratio of 1.2 in rats at 1 h post-dosing. The length of the aryl side chain was also investigated, and the benzyl length proved to be optimal as direct aryl attachment and phenethyl substituents significantly decrease A_{2A} activity. As with the previous series, the furan was replaced with various

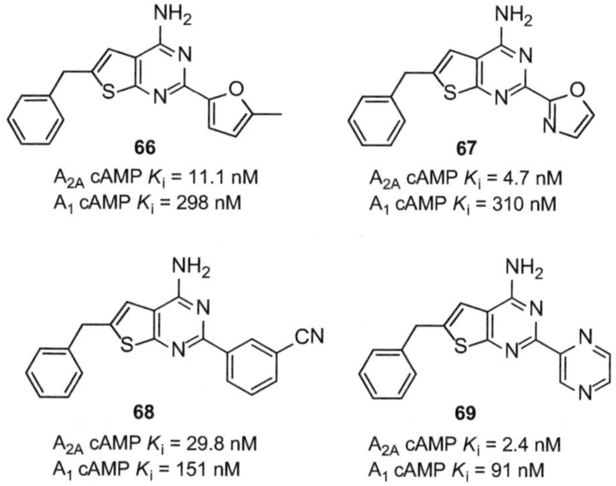

Fig. 20 Benzyl substituted thieno[2,3-*d*]pyrimidines

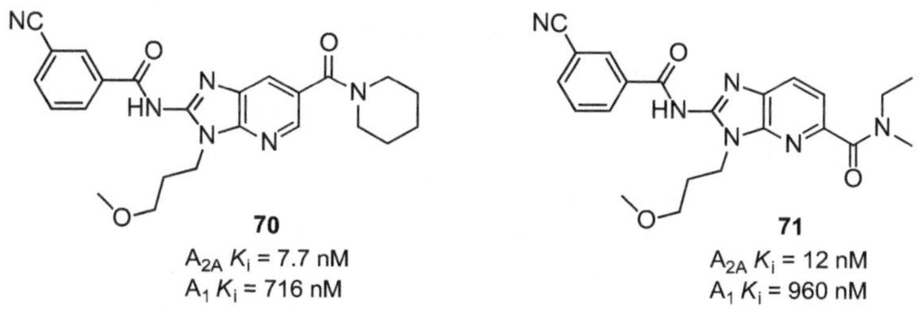

Fig. 21 2-Aminoimidazopyridines

5-membered heterocycles, but unlike the previous series, several oxazole and thiazole analogs maintained good in vitro activity particularly compound **67**. Unfortunately **67** was not active in mouse catalepsy at 3 mg/kg p.o. nor were any other 5-membered heterocyclic compounds that had good in vitro potency. Like the previous series, the 3-CN substituted phenyl was the ideal furan replacement, and compound **68** had good activity for A_{2A} and effectively reversed haloperidol-induced catalepsy in mice with an ED_{50} of <1.0 mg/kg p.o. Pyrazine was another furan replacement that maintained potent A_{2A} activity in compound **69**, but no in vivo efficacy or PK was reported. Substitution on the benzyl ring was investigated with this series of compounds and only *ortho* substituents like methoxy, chloro, and fluoro were tolerated.

A series of 2-aminoimidazopyridines **70** and **71** resulted from the optimization of a high-throughput screening hit (Fig. 21) [113]. Compound **70** showed good binding affinity for A_{2A} with 93-fold selectivity over A_1. A variety of amides were explored in the 6-position but generally suffered from poor microsomal stability. Moving the amide substituent to the 5-position resulted in compounds with similar binding affinities, but in general had better microsomal stability. Compound **71** was

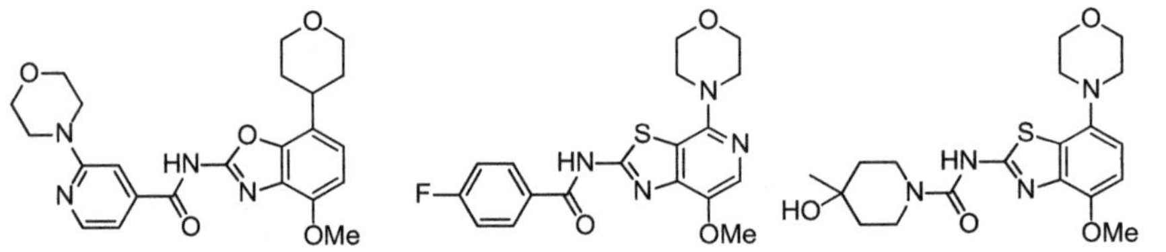

72
A_{2A} K_i = 25 nM
A_1 K_i = 8775 nM

73
A_{2A} K_i = 5 nM
A_1 K_i = 2100 nM

74 (SYN-115)
A_{2A} K_i = 8.3 nM
A_1 K_i = 2240 nM

Fig. 22 Substituted 4-morpholino-benzothiazoles

a potent binder for A_{2A} and showed moderate plasma and brain exposures when dosed orally, but were not evaluated in animal models of PD.

Benzoxazole-, thiazolopyridine-, and benzothiazole-based series of compounds have been reported in the patent literature by Hoffmann-La Roche [114–116]. Compounds **72** and **73** are representative examples of the benzoxazole and thiazolopyridine series, respectively, but no efficacy in PD models was reported (Fig. 22) [117, 118]. Further optimization of this scaffold led to the urea compound **74** (tozadenant, SYN115) which had good potency for A_{2A} and was 270-fold and >4,000-fold selective versus A_1 in binding and functional assays, respectively. Tozadenant had an ED_{50} of 0.5 mg/kg when dosed orally in the (2-[(2-aminoethylamino)carbonylethyl-phenyl-ethylamino]-5′-ethylcarboxamido-adenosine) (APEC)-induced hypolocomotion model in rats [50, 119]. The compound also showed good PK parameters in both rats and dogs. In January 2007, Synosia acquired the rights from Roche to develop tozadenant for the treatment of PD [120]. Tozadenant recently completed a phase II/III clinical trials for PD (clinicaltrials.gov identifier: NCT01283594) and for a phase 0 clinical trial for cocaine dependence (clinicaltrials.gov identifier: NCT00783276). In February 2011, Synosia was acquired by Biotie Therapeutics. More recently Biotie has granted UCB Pharma a license for exclusive worldwide rights to tozadenant where they will be responsible for the ongoing/additional phase III trials and commercialization [121].

Domain Therapeutics is evaluating a selective A_{2A} antagonist DT1133, but no structure has been disclosed.[1] A recent patent application from the company claims a series of substituted imidazo[1,2-*a*]pyridines as A_{2A} antagonists (Fig. 23) [122]. In general, compounds had good binding affinity, but no A_1 activity was reported. Compound **75** had good affinity for A_{2A} and was able to reverse haloperidol-induced catalepsy at 10 mg/kg p.o. The related indazole **76** and a benzoxazole **77** were both very potent binders for A_{2A}, but no PK or in vivo data was reported for these compounds.

A series of benzofurans from Kissei Pharmaceuticals has been reported in a patent application (Fig. 24) [123]. Unfortunately, biological data was only reported

[1] DT-1133 was listed on the Doamin Therapeutics website http://www.domaintherapeutics.com/.

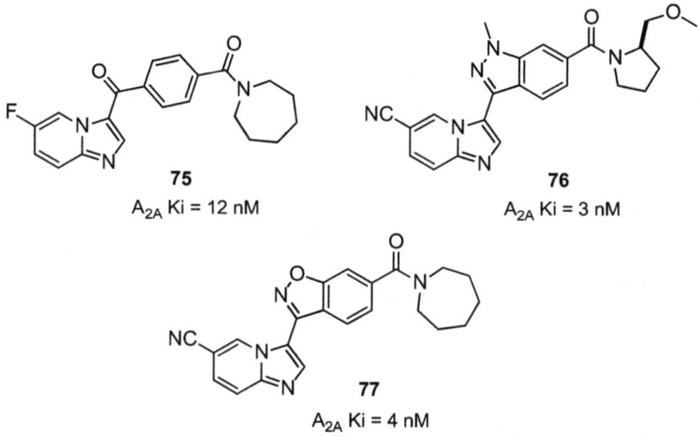

Fig. 23 Imidazopyridines

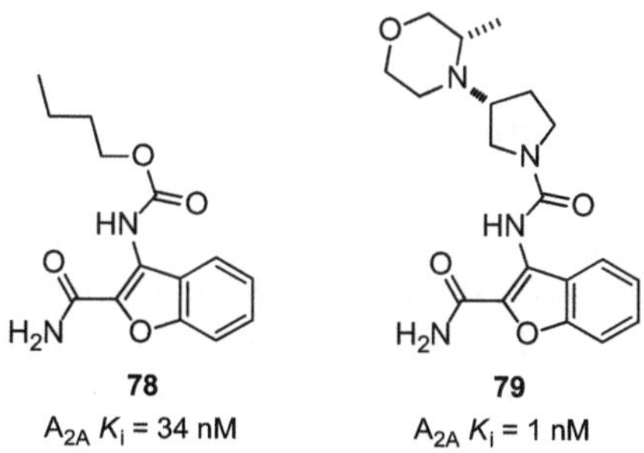

Fig. 24 Substituted benzofurans

for a limited number of compounds. The amino group in the 3-position of the benzofuran was substituted by various carbamate and urea linkages. The carbamate **78** was the only compound, of this type, with reported A_{2A} activity. The urea **79** was the most potent of the analogs that had A_{2A} activity reported.

A series of 4-aryl substituted benzofurans was identified and the 4-phenyl substituent was found to be the most consistent group for A_{2A} activity (Fig. 25) [124]. The morpholine amide **80** showed good affinity for A_{2A} and significantly reversed (78%) CGS-induced catalepsy in mice at 10 mg/kg p.o. Further exploration of this scaffold showed that the methyl carbamate **81** also had good A_{2A} activity and reversed catalepsy by 86% at 10 mg/kg p.o. A variety of amides was prepared from a related 4-morpholinobenzofuran series and were evaluated in vitro and in vivo [125]. The analogs in general exhibited similar in vitro potencies with respect to the 4-phenylbenzofurans, but also had increased aqueous solubility and improved PK properties. Compound **82** was identified having good in vitro activity for A_{2A} and excellent activity reversing (76% at 10 mg/kg p.o.) CGS-induced

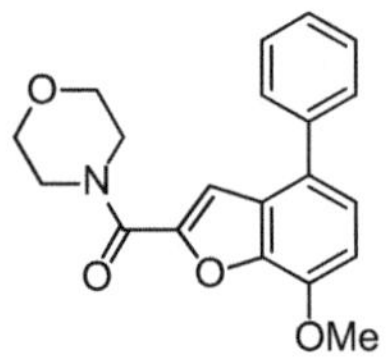

80

A_{2A} % inh at 10 nM = 39%
A_{2A} % inh at 100 nM = 81%
A_1 % inh at 1 μM = 37%
A_1 % inh at 10 μM = 82%

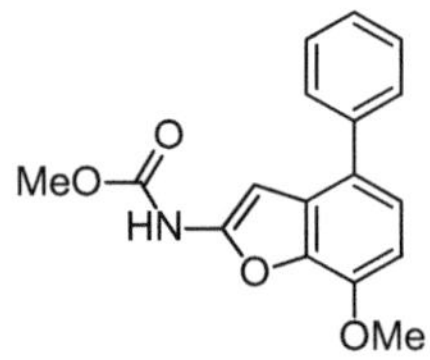

81

A_{2A} % inh at 10 nM = 28%
A_{2A} % inh at 100 nM = 85%

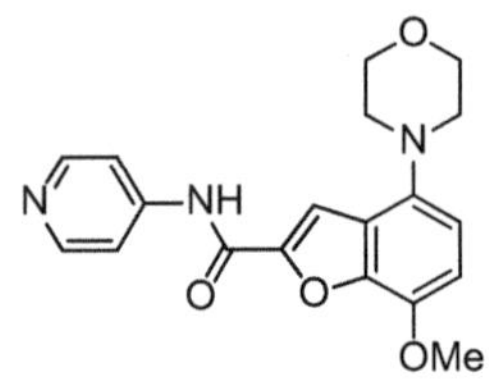

82

A_{2A} % inh at 10 nM = 30%
A_{2A} % inh at 100 nM = 73%

Fig. 25 4-Aryl and 4-morpholino substituted benzofurans

catalepsy in mice. The compound also showed significant reduction in motor disabilities and increased locomotor activity in MPTP-treated marmosets for 5 h after an oral dose of 10 mg/kg.

5 Monocyclic A_{2A} Antagonists

An aminopyrazine-based series of compounds was reported as dual A_{2A}/A_1 receptor antagonists (Fig. 26) [126]. Compound **83** had potent affinity for both A_{2A} and A_1 and significantly reversed haloperidol-induced catalepsy in mice at 3.2 mg/kg p.o. Further optimization of the pyrazine scaffold led to the identification of **84** (ASP-5854) that has been characterized extensively in numerous animal models of PD and cognition [127]. ASP-5854 reversed CGS-21680-induced and haloperidol-induced catalepsy in mice with ED_{50} values of 0.15 and 0.07 mg/kg p.o., respectively, and had a minimum effective dose of 0.1 mg/kg to reverse haloperidol-induced catalepsy in rats. In rhesus monkeys, ASP-5854 gave >85% occupancy of A_{2A} receptors in the striatum and reversed haloperidol-induced catalepsy with an ED_{50} of 0.1 mg/kg p.o [128]. ASP-5854 effectively potentiated L-DOPA-induced rotation in 6-OHDA-lesioned rats with a minimum effective dose of 0.03 mg/kg p.o. Statistical significance was reached in MPTP-treated mice and MPTP-treated marmoset models at 0.1 and 1 mg/kg p.o, respectively [129]. ASP-5854 produced positive results in scopolamine-induced memory deficits in the mouse Y-maze test and rat passive avoidance test, both models of cognition, with minimum effective doses of 0.3 and 0.1 mg/kg p.o., respectively. The selective A_{2A} antagonist KW-6002 (**6**) was also tested in parallel for comparison in these models and was significantly less effective having a minimum effective dose of 10 mg/kg p.o. in the mouse Y-maze test and no effect in the rat passive avoidance test up to 10 mg/kg p.o. These results suggest that a dual A_{2A}/A_1 may provide added benefit to PD patients to address the cognitive impairments associated with the disease.

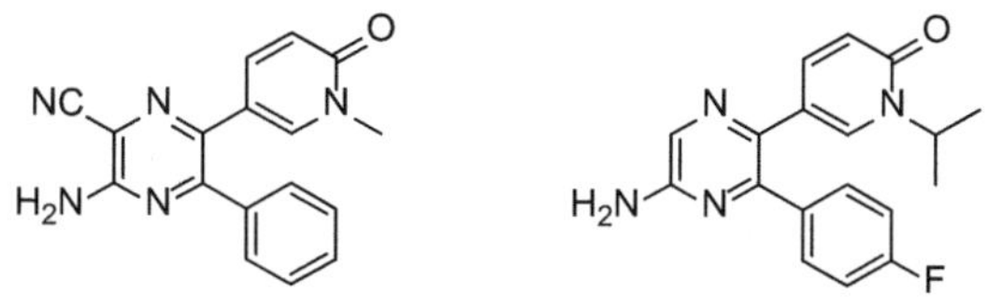

83
A_{2A} K_i = 2.4 nM
A_1 K_i = 22.5 nM

84 (ASP-5854)
A_{2A} K_i = 1.8 nM
A_1 K_i = 9.0 nM

Fig. 26 Pyridone substituted pyrazines

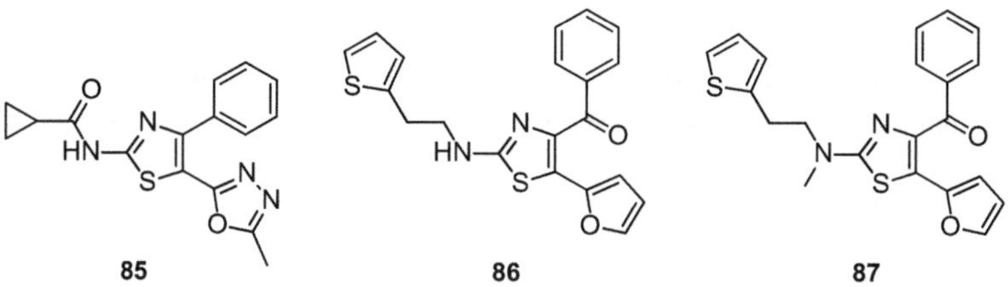

85
A_{2A} K_i = 28 nM
A_1 K_i = 150 nM

86
A_{2A} K_i = 12 nM
A_1 K_i > 6700 nM

87
A_{2A} K_i = 8170 nM
A_1 K_i > 10000 nM

Fig. 27 Heterocyclic substituted 2-amino-thiazoles

A series of carboxamides of 2-amino-4-phenyl-thiazoles were prepared as adenosine A_{2A} receptor antagonists (Fig. 27) [130]. The optimization of this scaffold identified that the cyclopropyl amide was the optimal amide substituent. Further evaluation into the substitution at the 5-position of the thiazole led to the 5-methyl-[1,2,4]-oxadiazole compound **85** which had good affinity for A_{2A} but very modest selectivity versus A_1. Compound **85** also showed good oral activity in the haloperidol-induced hypolocomotion model in mice with an ED_{50} of 7 mg/kg. A series of related substituted 2-amino-5-benzoyl-4-(2-furyl)thiazoles were designed from a high-throughput screening hit [131]. Lead optimization led to compound **86** which had good affinity for A_{2A} and excellent selectivity versus A_1. Interestingly, a simple methyl substitution on the nitrogen (**87**) dramatically decreased the binding affinity for A_{2A}. No in vivo evaluation of **86** was reported.

A series of substituted 1,2,4-triazines were recently reported as potent A_{2A} antagonists (Fig. 28) [132]. The 5-position was substituted with a variety of phenyl and 4-pyridyl groups that had good A_{2A} activity but generally had low selectivity against A_1. They were claimed as dual A_{2A}/A_1 antagonists in a recent patent application [133]. Compound **88** had good oral exposure and bioavailability in rats and had a brain-to-plasma ratio of 3.2 at 0.5 h. This compound was able to reverse haloperidol-induced catalepsy in rats at 1 and 2 h with an $ED_{50} = 0.2$ mg/kg p.o. for both time points. A co-crystal structure was obtained with **88**, and it showed hydrogen bonding of the pyridyl nitrogen with water molecules and that generated a second series of compounds that replace the pyridine with cyclic amines [134]. Compound **89** and **90** are representative compounds containing a cyclic amine, and they each show increased selectivity versus A_1. The hydrogen bonding

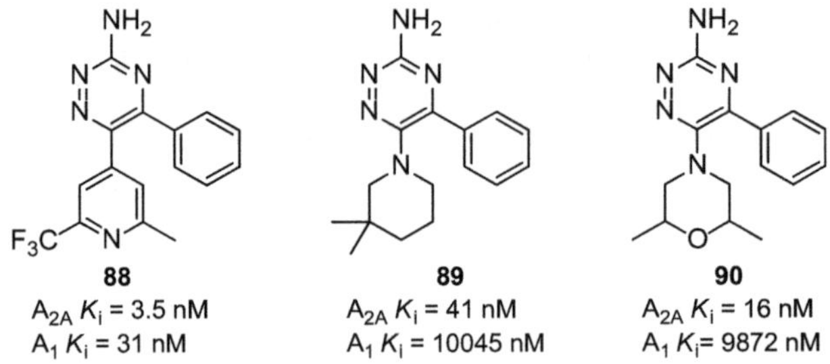

Fig. 28 Substituted 1,2,4-triazines

of the morpholine oxygen is presumed to be the reason for higher affinity compared to the piperidine compound **89**. No data was reported for their activity in animal models of PD.

Palobiofarma is currently investigating an A_{2A} antagonist PBF509 (structure not disclosed) for the treatment of PD.[2] A phase I trial to assess safety and tolerability has been completed (clinicaltrials.gov identifier: NCT01691924). A recent patent application claimed a series of substituted aminopyrimidines (Fig. 29) [135]. Limited biological data was reported for a select amount of compounds including **91** which has good binding affinity for A_{2A} and also good functional activity. A variety of substituents were explored in the 6-position, and it was shown that the pyrazine and thiazole analogs **92** and **93**, respectively, were two of the most potent compounds reported. No in vivo activity or PK data was given.

A series of trisubstituted pyrimidines was reported to have potent A_{2A} activity (Fig. 30) [136]. Compound **94** was a potent A_{2A} and A_1antagonist, but it had poor metabolic stability that was attributed to the unsubstituted furan moiety. SAR studies resulted in the synthesis of the thiazole compounds **95** and **96** that maintained potency for A_{2A}. Interestingly, replacing the benzylic methylene, present in **95**, with an oxygen atom (**96**) increased the selectivity against A_1. Compound **96** was further optimized by replacing the thiazole with the dimethylpyrazole to give compound **97** that was very potent for A_{2A} and had excellent selectivity versus A_1. Despite the increase in selectivity, **97** suffered from poor solubility, so basic amines were appended to the phenol ring to increase solubility and improve PK properties. A number of the amino alkyl compounds had good binding affinities for A_{2A} and good selectivity versus A_1, but had significant hERG liabilities. Compound **98** had the best hERG profile (patch clamp $IC_{50} = 1.2\ \mu M$) and showed oral activity in the haloperidol-induced catalepsy in rats with a minimal effective dose of 10 mg/kg. Despite very high clearance in rats, **98** exhibited desirable brain levels up to 4 h post-dosing. Further optimization of this scaffold led to compound

[2] In June, 2012, on the pipeline website, PBF509 was listed as being in phase I development as an adenosine A_{2A} antagonist for the potential treatment of Parkinson's disease. http://www.palobiofarma.com.

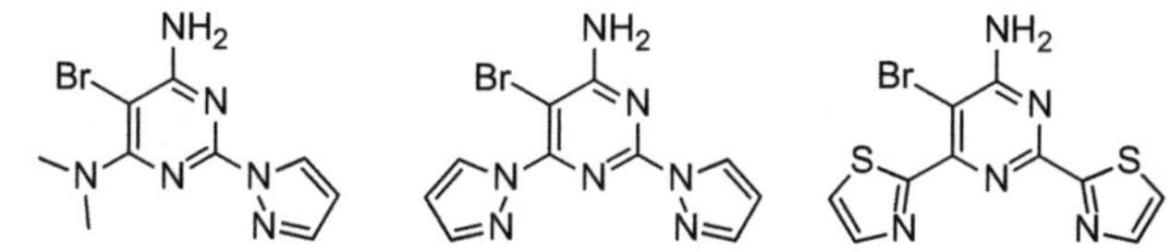

91 A_{2A} K_i = 35 nM; A_{2A} cAMP K_i = 50 nM

92 A_{2A} K_i = 12 nM; A_{2A} cAMP K_i = 25 nM

93 A_{2A} K_i = 1 nM; A_{2A} cAMP K_i = 12 nM

Fig. 29 Substituted bromoaminopyrimidines

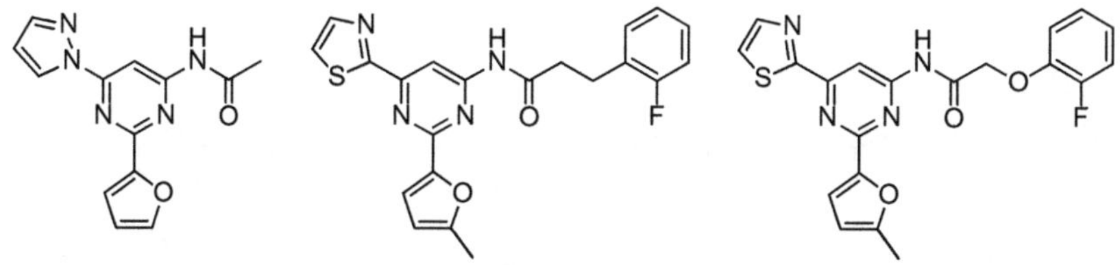

94 A_{2A} K_i = 0.6 nM; A_1 K_i = 9.2 nM

95 A_{2A} K_i = 16 nM; A_1 K_i = 410 nM

96 A_{2A} K_i = 7.5 nM; A_1 K_i = 650 nM

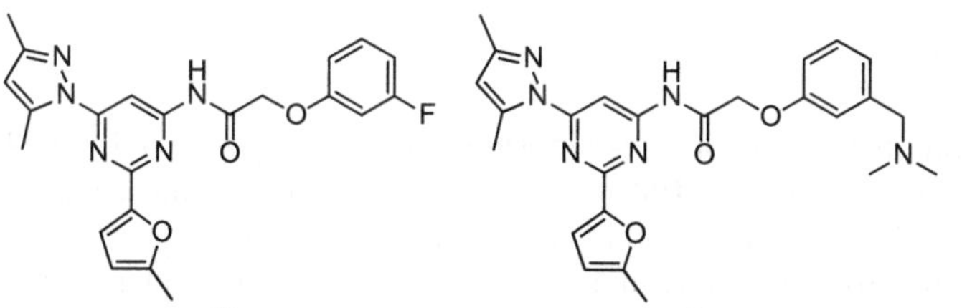

97 A_{2A} K_i = 3.7 nM; A_1 K_i = 9700 nM

98 A_{2A} K_i = 0.4 nM; A_1 K_i = 60 nM

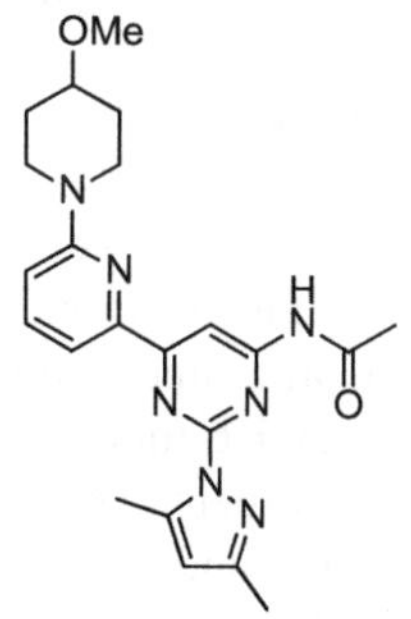

99 A_{2A} K_i = 0.4 nM; A_1 K_i = 77 nM

Fig. 30 Various trisubstituted pyrimidines

99, which had potent affinity for A_{2A} and A_1. Compound **99** effectively reversed catalepsy in rats at 1 mg/kg p.o. and had a brain-to-plasma ratio of ~4 [137]. This compound also potentiated L-DOPA-induced rotation in the 6-OHDA lesion model in rats with a minimum effective dose of 3 mg/kg p.o. 4-Methoxypiperidine was replaced with a number of cyclic amines to increase solubility, but generally resulted in decreased A_{2A} activity and decreased selectivity against A_1 [138].

A series of 2-amino-*N*-pyrimidin-4-ylacetamides were prepared with compound **69** being the representative example (Fig. 31) [139]. Compound **100** had a high affinity for A_{2A} and was selective versus A_1. The compound exhibited very good exposure (2,700 ng/g) in rat brain at 2 h after a 10 mg/kg oral dose. The compound

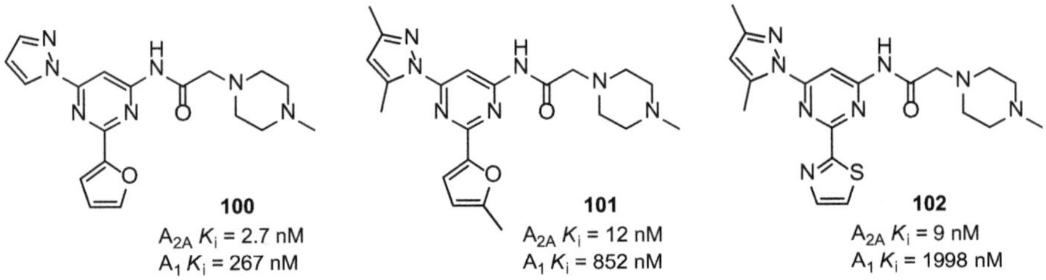

Fig. 31 Piperazine substituted pyrimidine acetamides

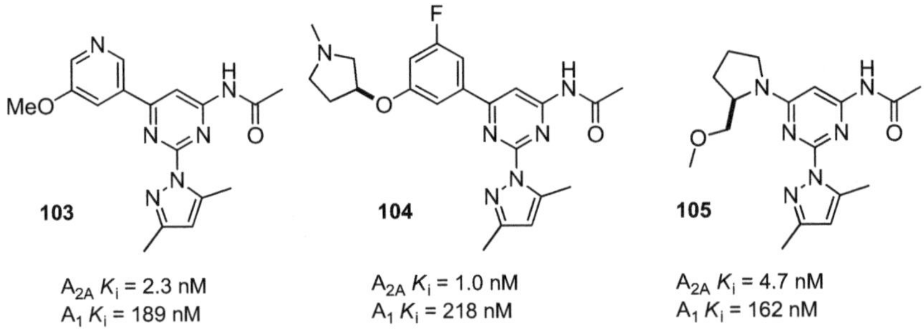

Fig. 32 Acyl amino pyrimidines

effectively reversed haloperidol-induced catalepsy in rats at 10 mg/kg p.o. Like **94**, the unsubstituted furan in compound **100** was being metabolized, so there was an effort to find a suitable furan replacement [140]. Several combinations consisting of 5- and 6-membered heterocycles were made to replace the furan and pyrazole substituents. Compounds **101** and **102** were identified as the most promising compounds which have good binding affinities and good selectivity versus A_1. Further evaluation of **101** showed it to be a potent inhibitor of hERG with and IC_{50} of 950 nM (patch clamp). The basicity of the piperazine was enhancing the hERG activity, so various replacements were prepared in effort to decrease the hERG liability. Although the hERG inhibition could not be completely eliminated, the removal of one of the basic amines decreased the hERG activity. Several compounds within the series showed activity in haloperidol-induced catalepsy in rats.

A related series of 2,6-diaryl-4-acylaminopyrimidines were developed with improved solubility and metabolic stability (Fig. 32) [141]. Compound **103** was a potent A_{2A} antagonist that had good plasma exposure in rats and reversed catalepsy at a dose of 3 mg/kg p.o. Unfortunately, **103** had poor solubility and low exposures in both dogs and monkeys. Compound **104** had high affinity for A_{2A}, was 218-fold selective against A_1, and was significantly more soluble than **103**. More importantly, the addition of the fluorine atom significantly increased metabolic stability, decreased intrinsic clearance, and reduced CYP inhibition compared to the corresponding *des*-fluoro analog, not shown. Further characterization of **104** showed that it significantly reversed haloperidol-induced catalepsy in rats at

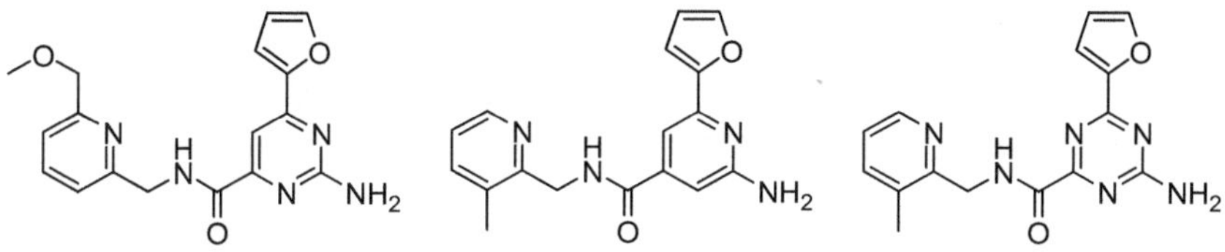

106
A_{2A} K_i = 1.7 nM
A_1 K_i = 43 nM

107
A_{2A} K_i = 126 nM
A_1 K_i = 921 nM

108
A_{2A} K_i = 18.6 nM
A_1 K_i = 1112 nM

Fig. 33 Pyrimidine, pyridine, and triazine carboxamides

30 mg/kg p.o. The direct attachment of the pyrrolidine to the pyrimidine core gave **105** which has the ability to reverse catalepsy in rats at 10 mg/kg p.o. Compound **105** was able to potentiate L-DOPA-induced rotational behavior in the 6-OHDA-lesioned rat at 10 and 30 mg/kg oral doses.

Analogs of pyrimidine-4-carboxamides have been reported to display good potency for A_{2A} and with moderate to good selectivity over other adenosine receptor subtypes (Fig. 33) [142]. Compound **106** is a representative example having binding affinities of 1.7, 43, 460, 1,740 nM for A_{2A}, A_1, A_{2B}, A_3, respectively. Compound **106** also had a minimum effective dose of 0.1 mg/kg in the haloperidol-induced hypolocomotion model in mice. This compound had very good bioavailability ($F = 90\%$) in rats with very good brain uptake. Replacing the pyridyl with a variety of substituted aryls generally gave compounds with good affinity for A_{2A} but with very modest selectivity against A_1. Related pyridine **107** and triazine **108** have also been reported, but are significantly less potent than the corresponding pyrimidines [143].

6 Radiolabeled A_{2A} Antagonists

Several A_{2A} antagonists have been radiolabeled to study binding affinities and receptor occupancy in an effort to better understand the pharmacology of A_{2A} receptor and its relevance in treating PD patients. There have been many efforts to identify radio and PET ligands that bind specifically to A_{2A} receptors in the brain without the complication of nonspecific binding. The first generation of PET ligands used the xanthine core as a starting point and eventually discovered [^{11}C] TMSX (**109**) which was used to map A_{2A} receptors in the brains of PD patients (Fig. 34) [144, 145]. Like the parent compound, **109** is susceptible to photoisomerization which prompted efforts to identify non-xanthine-based PET tracers.

Several studies were reported using [^{125}I]ZM241385 and [^{3}H]ZM241385 (structures not shown), but their use has been limited because of its good affinity for A_{2B} ($K_i = 75$ nM). The first non-xanthine tracer to be used as a versatile research tool was [^{3}H]SCH58261 (**110**) which showed reversible binding in rat striatal membranes [146]. This tracer has been used in numerous studies including A_{2A} receptor

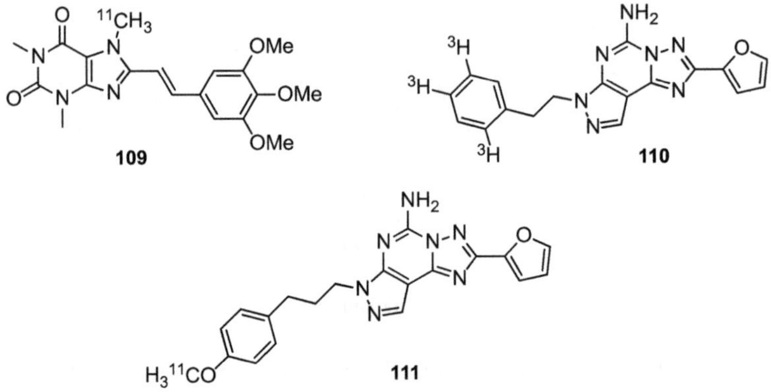

109 110 111

Fig. 34 Radioligands

labeling in human platelets and neutrophil membranes and autoradiography studies in rats [147–150].

Perhaps the most significant and best radioligand is [^{11}C]SCH-442416 (**111**) which has high affinity for A_{2A} (h$K_i = 0.05$ nM) and is very selective against other receptor subtypes (A_1 h$K_i = 1{,}111$ nM, A_{2B} h$K_i > 10{,}000$ nM, A_3 h$K_i > 10{,}000$ nM) [151]. It also has good potency and selectivity in rat A_{2A} r$K_i = 0.5$ nM and A_1 r$K_i = 1{,}815$ nM. Compound **111** had good brain exposure and had a brain distribution that was four to fivefold higher in the striatum compared to other brain compartments 15 min post-dosing. This compound was successfully used as a PET tracer in healthy male volunteers to study the relationships between dose, plasma exposure, and receptor occupancy of vipadenant **57** [152].

7 Dual A_{2A} and MAO-B Inhibitors

A series of benzothiazinones and thienothiazinones have been reported as dual A_{2A}/MAO-B inhibitors (Fig. 35) [38, 39]. The activity and expression levels of MAO-B in the human brain increase with age while MAO-A remains constant [40]. Under L-DOPA treatment, the increased MAO-B activity is associated with increased hydrogen peroxide formation, which could contribute to neurodegeneration related to PD. As mentioned previously, it is known that inhibiting MAO-B improves symptoms of PD, and two MAO-B inhibitors, selegiline and rasagiline, are currently approved and used for treatment of PD patients as monotherapy or in combination with nn-DOPA [42, 43]. Therefore it was proposed that a compound having dual action, A_{2A} antagonist and MAO-B inhibitor, may have improved therapeutic and neuroprotective potential in the treatment of PD. The benzothiazinone **112** has good affinity for A_{2A} and is moderately selective versus A_1 [38, 39]. This compound also has good potency for MAO-B, one of the

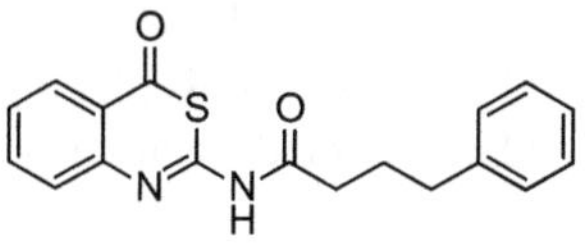

112
A_{2A} K_i = 39.5 nM
A_1 K_i = 2489 nM
MAO-B IC_{50} = 34.9 nM

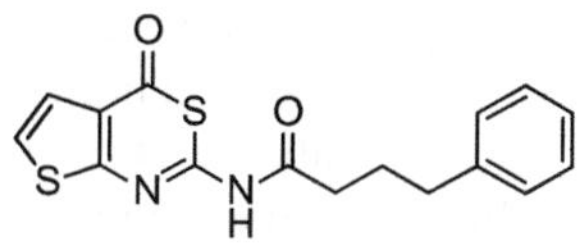

113
A_{2A} K_i = 82.5 nM
A_1 K_i = 330 nM
MAO-B IC_{50} = 69.7 nM

Fig. 35 Benzothiazinones and thienothiazinones

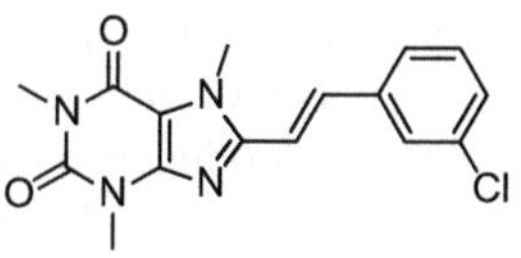

4 (CSC)
A_{2A} K_i = 54 nM
A_1 K_i = 28000 nM
MAO-B IC_{50} = 18.1

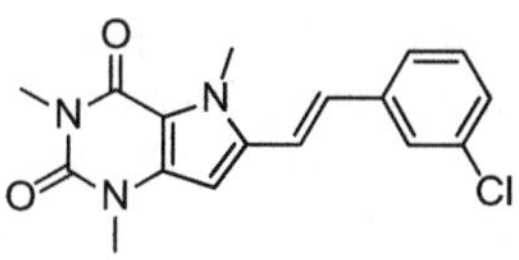

114
A_{2A} K_i = 260 nM
A_1 K_i = ND
MAO-B IC_{50} = 200 nM

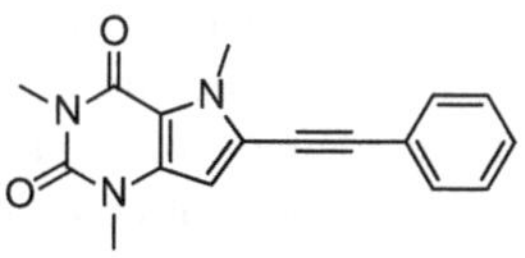

115
A_{2A} K_i > 10000 nM
A_1 K_i = ND
MAO-B IC_{50} = 173 nM

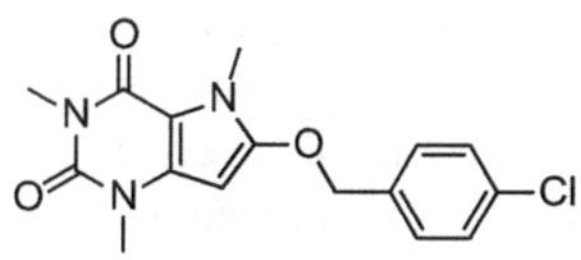

116
A_{2A} K_i > 10000 nM
A_1 K_i = ND
MAO-B IC_{50} = 48 nM

Fig. 36 9-Deazaxanthines

most potent from this series, and is quite selective versus MAO-A. The length of the amide side chain was optimal, as in **112**, and substitution on the phenyl ring generally resulted in decreased A_{2A} activity. The related thienothiazinone **113** was approximately twofold less potent for both A_{2A} and MAO-B, but the biggest difference between **112** and **113** is that **113** is significantly less selective (<5-fold) against A_1 compared to **112** (>60-fold). No in vivo data was reported for either series of compounds.

CSC (**4**, Fig. 1) is a potent and selective A_{2A} antagonist from the xanthine series of compounds; it is also known to be a potent MAO-B inhibitor [39, 55] A recent series of 9-deazaxanthine derivatives have been reported as dual A_{2A} antagonists/ MAO-B inhibitors with **114** being the most interesting of compounds (Fig. 36) [39]. It has a good balance of A_{2A} and MAO-B activity and was shown to dose-dependently reverse haloperidol-induced catalepsy with a minimum effective dose of 30 mg/kg. Replacement of the chlorine with substituents like CF_3, OMe, and Br resulted in slight decreases in both A_{2A} and MAO-B activities. Replacing the olefin side chain with and alkyne (**115**) or alkoxyl (**116**) linker dramatically decreased A_{2A} activity but maintained comparable MAO-B potency.

8 A_{2A} Antagonists with $mGlu_4$ PAMs

Metabotropic glutamate receptor-4 ($mGlu_4$) positive allosteric modulators (PAM) have been reported to produce antiparkinsonian-like effects in preclinical models of PD [153]. It has also been reported that co-administration of a group III mGlu agonist with an A_{2A} antagonist produced enhanced reversal of haloperidol-induced catalepsy that was significantly better than the activity produced by either agent when dosed alone [47]. More recently it was shown that co-administration of VU0364770, an $mGlu_4$ PAM, significantly increased preladenant's (**14**) ability to reverse catalepsy in rats. VU0364770 was able to reverse catalepsy alone at a subcutaneous dose of 30 mg/kg but not at 10 mg/kg s.c [45]. In this study, it was also assessed that preladenant was able to reverse catalepsy in rats at 0.3 mg/kg p.o., but not at 0.1 mg/kg. It was shown that VU0364770 (10 mg/kg s.c.) given in combination with preladenant (0.1 mg/kg p.o.), not an effective dose of either by itself, produced significant reversal of catalepsy. Also, the dosing of VU0364770 increased the amount of reversal for all doses of preladenant (0.3 and 1 mg/k p.o.) compared to when preladenant was dosed alone. Although not completely understood, the observed pharmacologic interaction of these two receptors suggests that synergy may be occurring on the same and/or different neuronal pathways. There are two striatal GABAergic projection pathways, the direct pathway that projects directly to the substantia nigra pars reticulata and the internal globus pallidus, and the indirect pathway that projects from the striatum to the external globus pallidus [46]. Loss of striatal dopaminergic innervation results in overactivation of the indirect pathway which contributes to the motor deficits observed in PD. Adenosine A_{2A} and $mGlu_4$ receptors both act on the indirect pathway, and inhibiting A_{2A} receptors or activating $mGlu_4$ receptors has been shown to reverse motor impairments in animal models of PD. The roles of these receptors on the indirect pathway seem to complement one another and could thus explain the observed pharmacology and also suggest that targeting these two receptors could provide enhanced treatment for PD patients. It would be exciting to see if one could target both receptors within the same molecule.

9 Summary

Adenosine A_{2A} receptor antagonists continue to be a sought-after target for the treatment of PD. A variety of chemical scaffolds have been reported with interesting SAR trends that showed dramatic activity differences, both in vitro and in vivo, from subtle modifications. Despite the relentless research efforts to identify potent compounds, there still remains the challenge of achieving selectivity, solubility, and acceptable pharmacokinetic/pharmacodynamic properties to progress into the clinic. There have been a number of compounds that are in or have been in clinical trials including istradefylline (**6**), preladenant (**14**), vipadenant (**57**), ST-1535 (**59**), tozadenant (**74**), V81444, and PBF509. The progression through the clinic with positive trial results indicates the potential of A_{2A} antagonists to treat PD patients with potential advantages over the current standard treatments, but the lack of robust efficacy in phase III trials gives pause to whether this target is really beneficial to PD patients. The approval of istradefylline is exciting, but it will be interesting to monitor how PD patients will respond after continued treatment. Examples of dual A_{2A}/A_1 and A_{2A}/MAO-B were also described as having the potential to provide added benefit to PD patients, via A_1 antagonism or MAO-B inhibition. The dual activities, however, need to be tested in a clinical setting to completely understand possible synergies. Combination of A_{2A} antagonists with $mGlu_4$ PAMs is another interesting approach that will surely receive more interest and facilitate more studies due to the initial positive results in rodent models of PD. The ongoing clinical trials with tozadenant, V81444, and PBF509 will hopefully help validate this target to treat Parkinson's disease.

References

1. Druri AN, Szent-Gyorgy A (1929) The physiological activity of adenine compounds with especial reference to their action upon the mammalian heart. J Physiol (Lond) 68:213–237
2. Van Calker D, Muller M, Hamprecht B (1979) Adenosine regulates via two different types of receptors, the accumulation of cyclic AMP in cultured brain cells. J Neurochem 33:999–1005
3. Stiles GL (1992) Adenosine receptors. J Biol Chem 267:6451–6454
4. Latini S, Pedata F (2001) Adenosine in the central nervous system: release mechanisms and extracellular concentrations. J Neurochem 79:463–484
5. Dunwiddie TV, Masino SA (2001) The role and regulation of adenosine in the central nervous system. Annu Rev Neurosci 24:31–55
6. Rosin DL, Robeva A, Woodard RL, Guyenet PG, Linden J (1998) Immunohistochemical localization of adenosine A_{2A} receptors in rat central nervous system. J Comp Neurol 401:163–186
7. Fredholm BB, Svenningsson P (1998) Striatal adenosine A_{2A} receptors-where are they? What do they do? Comments. Trends Pharmacol Sci 19:46–47
8. Ishiwata K, Mishina M, Kimura Y, Oda K, Sasaki T, Ishii K (2005) First visualization of adenosine A_{2A} receptors in the human brain by positron emission tomography with [^{11}C] TMSX. Synapse 55:133–136

9. Svenningsson P, Hall H, Sedvall G, Fredholm BB (1997) Distribution of adenosine receptors in the postmortem human brain: an extended autoradiographic study. Synapse 27:322–335
10. Ledent C, Vaugeois JM, Schiffmann SN, Pedrazzini T, Wl Yacoubi M, Vanderhaeghen JJ, Constentin J, Heath JK, Vassart G, Parmentier M (1997) Aggressive Hypoalgesia, and high blood pressure in mice lacking the adenosine A_{2A} receptor. Nature 388:674–678
11. Chem Y, King K, Lai H (1992) Molecular cloning of a novel adenosine receptor gene from rat brain. Biochem Biophys Res Commun 185:304–309
12. Meng F, Xie G, Chalmers D, Morgan C, Watson SJ, Akil H (1994) Cloning and expression of the A_{2A} adenosine receptor from guinea pig brain. Neurochem Res 19:613–621
13. Furlong TJ, Pierce KD, Selbie LA, Shine J (1992) Molecular characterization of a human brain adenosine A_2 receptor. Brain Res 15:62–66
14. Fink JS, Weaver DR, Rivkees SA, Peterfreund RA, Pollack AE, Adler EM, Reppert SM (1992) Molecular cloning of the rat A_2 adenosine receptor: selective co-expression with D_2 dopamine receptors in rat striatum. Mol Brain Res 14:186–195
15. Schiffmann SN, Lipert F, Vassart G, Vanderhaeghen JJ (1991) Distribution of adenosine A_2receptor mRNA in the human brain. Neurosci Lett 130:177–181
16. Pollack AE, Fink JS (1995) Adenosine antagonists potentiate D2 dopamine dependent activation of Fos in the striatopallidal pathway. Neuroscience 68:721–728
17. Ungerstedt U (1968) 6-Hydroxydopamine-induced generation of central monoamine neurons. Eur J Pharmacol 5:107–110
18. Ungerstedt U, Arbuthnott GW (1970) Quantitative recording of rotational behavior in rats after 6-hydroxy-dopamine lesions of the nigrostriatal dopamine system. Brain Res 24:485–493
19. Luthman J, Fredriksson A, Sundstroem E, Jonsson G, Archer T (1989) Selective lesion of central dopamine or noradrenaline neuron systems in the neonatal rat: motor behavior and monoamine alteration at adult stage. Behav Brain Res 33:267–277
20. Bankiewicz KS (1991) MPTP-induced parkinsonism in nonhuman primates. Methods Neurosci 7:168–182
21. Jakowec MW, Petzinger GM (2004) 1-methyl-4-phenyl-1,2,3,6-tetrahydropyridine-induced lesion model of Parkinson's disease, with emphasis on mice and nonhuman primates. Comp Med 54:497–513
22. Campos-Romo A, Ojeda-Flores R, Moreno-Briseno P, Fernandez-Ruiz J (2009) Quantitative evaluation of MPTP-treated nonhuman parkinsonian primates in the hallway task. J Neurosci Methods 177:361–368
23. Chen JF, Xu K, Petzer JP, Staal R, Xu YH, Beilstein M, Sonsalla PK, Castagnoli K, Castagnoli N, Schwarzchild MA (2001) Neuroprotection by caffeine and A_{2A} adenosine receptor inactivation in a model of Parkinson's disease. J Neurosci 21:RC143/1–RC143/6
24. Grondin R, Bedard PJ, Hadj Tahar A, Gregoire L, Mori A, Kase H (1999) Antiparkinsonian effect of a new adenosine A_{2A} receptor antagonist in MPTP-treated monkeys. Neurology 52:1673–1677
25. Ongini E, Monopoli A, Impagnatiello F, Fredduzzi S, Schwarzchild M, Chen JF (2001) Dual actions of A_{2A} adenosine receptor antagonists on motor dysfunction and neurodegenerative processes. Drug Dev Res 52:379–386
26. Ikeda K, Kurokawa M, Aoyama S, Kuwana Y (2002) Neuroprotection by adenosine A_{2A} receptor blockade in experimental models of Parkinson's disease. J Neurochem 80:262–270
27. Kanda T, Jackson MJ, Smith LA, Pearce RKB, Nakamura J, Kase H, Kuwana Y, Jenner P (1998) Adenosine A_{2A} antagonist: a novel antiparkinsonian agent that does not provoke dyskinesia in parkinsonian monkeys. Ann Neurol 43:507–513
28. Koga K, Kurokawa M, Ochi M, Nakamura J, Kuwana Y (2000) Adenosine A_{2A} receptor antagonists KF17837 and KW-6002 potentiate rotation induced by dopaminergic drugs in hemi-Parkinsonian rats. Eur J Pharmacol 408:249–255
29. Kanda T, Jackson MJ, Smith LA, Pearce RK, Nakamura J, Kase H, Kuwana Y, Jenner P (2000) Combined use of the adenosine A_{2A} antagonist KW-6002 with L-DOPA or with

selective D1 or D2 dopamine agonists increases antiparkinsonian activity but not dyskinesia in MPTP-treated monkeys. Exp Neurol 162:321–327
30. Hauser RA, Hubble JP, Truong DD (2003) Randomized trial of adenosine A_{2A} receptor antagonist istradefylline in advanced PD. Neurology 61:297–303
31. Bara-Jimenez W, Sherzai A, Dimitrova T, Favit A, Bibbiani F, Gillespie M, Morris MJ, Mouradian MM, Chase TN (2003) Adenosine A_{2A} receptor antagonist treatment of Parkinson's disease. Neurology 61:293–296
32. Ballarin M, Reiriz J, Ambrosio S, Mahy N (1995) Effect of locally infused 2-cholroadenosine, an A_1 receptor agonist, on spontaneous and evoked dopamine release in rat neostriatum. Neurosci Lett 185:29–32
33. Moore KA, Nicoll RA, Schmitz D (2003) Adenosine gates synaptic plasticity at hippocampal mossy fiber synapses. Proc Natl Acad Sci U S A 100:14397–14402
34. Rebola N, Pinheiro PC, Oliveira CR, Malva JO, Cunha RA (2003) Subcellular localization of adenosine A_1 receptors in nerve terminal and synapses of the rat hippocampus. Brain Res 987:49–58
35. Maemoto T, Miho T, Takuma M, Noriko U, Hideaki M, Katsuya H, Takayuki Y, Kiyoharu S, Satoru K, Atsushi A, Akinori I, Nobuya M, Seitaro M (2004) Pharmacological characterization of FR194921, a new potent, selective, orally active antagonist for central adenosine A_1 receptors. J Pharmacol Sci 96:42–52
36. Gottlieb SS, Skettino SL, Wolff W, Beckman E, Fisher ML, Freudenberger R, Gladwell T, Marshall J, Cines M, Bennett D, Liittschwager EB (2000) Effects of BG9719 (CVT-124), an A_1-adenosine receptor antagonist, and furosemide on glomerular filtration rate and natriuresis in patients with congestive heart failure. J Amer Coll Cardiol 35:56–59
37. Givertz MM, Massie BM, Fields TK, Pearson LL, Dittrich HC (2007) The effects of KW-3902, an adenosine A1-receptor antagonist, on diuresis and renal function in patients with acute decompensated heart failure and renal impairment or diuretic resistance. J Amer Coll Cardiol 50:1551–1560
38. Stobel A, Schlenk M, Hinz S, Kuppers P, Heer P, Gutschow M, Muller CE (2013) Dual targeting of adenosine A_{2A} receptors and monoamine oxidase B by 4*H*-3,1-benzothiazin-4-ones. J Med Chem 56:4580–4596
39. Rivara S, Piersanti G, Bartoccini F, Diamantini G, Pala D, Riccioni T, Stasi MA, Cabri W, Borsini F, Mor M, Tarzia G, Minetti P (2013) Synthesis of (*E*)-8-(3-chlorostyryl)caffeine analogues leading to 9-deazaxanthine derivatives as dual A_{2A} antagonists/MAO-B inhibitors. J Med Chem 56:1247–1261
40. Nicotra A, Pierucci F, Parvez H, Senatori O (2004) Monoamine oxidase expression during development and aging. Neurotoxicology 25:155–165
41. Sayre LM, Perry G, Smith MA (2008) Oxidative stress and neurotoxicity. Chem Res Toxicol 21:172–188
42. Palhagen S, Heinonen E, Hagglund J (2006) Selegiline slows the progression of the symptoms of Parkinson's disease. Neurology 66:1200–1206
43. Parkinson Study Group (2004) A controlled, randomized, delayed-start study of rasagiline in early Parkinson's disease. Arch Neurol 61:561–566
44. Schapira AH, Stocchi F, Borgohain R, Onofrj M, Bhatt M, Lorenzana P, Lucini V, Giuliani R, Anand R (2013) Long-term efficacy and safety of safinamide as add-on therapy in early Parkinson's disease. Eur J Neurol 20:271–280
45. Jones CK, Bubser M, Thompson AD, Dickerson JW, Turle-Lorenzo N, Amalric M, Blobaum AL, Bridges TM, Morrison RD, Jadhav S, Engers DW, Italiano K, Bode J, Daniels JS, Lindsley CW, Hopkins CR, Conn PJ, Niswender CM (2012) The metabotropic glutamate receptor 4-positve allosteric modulator VU0364770 produces efficacy alone and in combination with L-DOPA or an adenosine A_{2A} antagonist in preclinical rodent models of Parkinson's disease. J Pharmacol Exp Ther 340:404–421
46. Conn PJ, Barraglia G, Marino MJ, Nicoletti F (2005) Metabotropic glutamate receptors in the basal ganglia motor circuit. Nat Rev Neurosci 6:787–798

47. Lopez S, Turle-Lorenzo N, Archer F, De Leonibus E, Mele A, Amalric M (2007) Targeting group III metabotropic glutamate receptors produces complex behavioral effects in rodent models of Parkinson's disease. J Neurosci 27:6701–6711
48. de Lera Ruiz M, Lim Y-H, Zheng J (2013) Adenosine A_{2A} receptor as a drug discovery target. J Med Chem 57:3623–3650. doi:10.1021/jm4011669
49. Shook BC, Jackson PF (2011) Adenosine A_{2A} receptor antagonists and Parkinson's disease. ACS Chem Neurosci 2:555–567
50. Shah U, Hodgson R (2010) Recent progress in the discovery of adenosine A2A receptor antagonists for the treatment of Parkinson's disease. Curr Opin Drug Disc Dev 13:466–480
51. Pinna A (2009) Novel investigational adenosine A_{2A} receptor antagonists for Parkinson's disease. Expert Opin Investig Drugs 18:1619–1631
52. Baraldi PG, Tabrizi MA, Gessi S, Borea PA (2008) Adenosine receptor antagonists: translating medicinal chemistry and pharmacology into clinical utility. Chem Rev 108:238–263
53. Valls MD, Cronstein BN, Montesinos MC (2009) Adenosine receptor agonists for promotion of dermal would healing. Biochem Pharmacol 77:1117–1124
54. Fredholm BB (1980) Are methylxanthines effects due to antagonism of endogenous adenosine? Trends Pharmacol Sci 1:129–132
55. Jacobson KA, Gallo-Rodriguez C, Melman N, Fischer B, Maillard M, van Bergen A, van Galen PJ, Karton Y (1993) Structure-activity relationships of 8-styrylxanthines as A_2-selective adenosine antagonists. J Med Chem 36:1333–1342
56. Sauer R, Maurinsh J, Reith U, Fulle F, Klotz KN, Muller CE (2000) Water-soluble phosphate prodrugs of 1-propargyl-8-styrylxanthine derivatives, A_{2A} selective adenosine receptor antagonists. J Med Chem 43:440–448
57. Kase H (2003) The adenosine A_{2A} receptor selective antagonist KW6002: research toward a novel nondopaminergic therapy for Parkinson's disease. Neurology 61:S97–S100
58. Seale TW, Abla KA, Shamim MT, Carney JM, Daly JW (1988) 3,7-Dimethyl-1-propargylxanthine: a potent and selective in vivo antagonist of adenosine analogs. Life Sci 43:1671–1684
59. Hokemeyer J, Burbiel J, Muller CE (2004) Multigram-scale synthesis, stability and photoreactions of A_{2A} adenosine receptor antagonists with 8-styrylxanthine structure: potential drugs for Parkinson's disease. J Org Chem 69:3308–3318
60. Dungo R, Deeks ED (2013) Istradefylline: first global approval. Drugs 73:875–882
61. Kanda T, Tashiro T, Kuwana Y, Jenner P (1998) Adenosine A2A receptors modify motor function in MPTP-treated common marmosets. Neuroreport 9:2857–2860
62. Kyowa Hakko Kogyo Co. Ltd. (2008) Kyowa Hakko receives not approvable letter from FDA for istradefylline (KW-6002). Press Release, February 28
63. Williams M, Francis J, Ghai G, Braunwalder A, Psychoyos S, Stone GA, Cash WD (1987) Biochemical characterization of the triazoloquinazoline, CGS15943, a novel, non-xanthine adenosine antagonist. J Pharmacol Exp Ther 241:415–420
64. Sorges R, Howard HR, Browne RG, Lebel LA, Seymour PA, Koe BK (1990) 4-Amino[1,2,4] triazolo[4,3-a]quinoxalines. A novel class of potent adenosine receptor antagonists and potential rapid-onset antidepressants. J Med Chem 33:2240–2254
65. Gatta F, Del Giudice MR, Borioni A, Borea PA, Dionisotti S, Ongini E (1993) Synthesis of imidazo[1,2-*c*]pyrazolo[4,3-*e*]pyrimidines, pyrazolo[4,3-*e*]1,2,4-triazolo[1,5-*c*]pyrimidines and 1,2,4-triazolo[5,1-*i*]purines: new potent adenosine A2 receptor antagonists. Eur J Med Chem 28:569–576
66. Ongini E (1997) SCH 58261: a selective A_{2A} adenosine receptor antagonist. Drug Dev Res 42:63–70
67. Shah U, Boyle CD, Chackalamannil S, Neustadt BR, Lindo N, Greenlee WJ, Foster C, Arik L, Zhai Y, Ng K, Wang S, Monopoli A, Lachowicz JE (2008) Biaryl and heteroaryl derivatives of SCH 58261 as potent and selective adenosine A2A receptor antagonists. Bioorg Med Chem Lett 18:4199–4203

68. Baraldi PG, Cacciari B, Romagnoli R, Spalluto G, Monopoli A, Ongini E, Varani K, Borea PA (2002) 7-Substituted 5-amino-2-(2-furyl)pyrazolo[4,3-*e*]-1,2,4-triazolo[1,5-*c*]pyrimidines as A_{2A} adenosine receptor antagonists: a study on the importance of modifications at the side chain on the activity and solubility. J Med Chem 45:115–126
69. Neustadt BR, Hao J, Lindo N, Greenlee WJ, Stamford AW, Tulshian D, Ongini E, Hunter J, Monopoli A, Bertorelli R, Foster C, Arik L, Lachowicz J, Ng K, Feng KI (2007) Potent, selective, and orally active adenosine A2A receptor antagonists: arylpiperazine derivatives of pyrazolo[4,3-e]-1,2,4-triazolo[1,5-c]pyrimidines. Bioorg Med Chem Lett 17:1376–1380
70. Hodgson RA, Bertorelli R, Varty GB, Lachowicz JE, Forlani A, Fredduzzi S, Cohen-Williams ME, Higgins GA, Impagnatiello F, Nicolussi E, Parra LE, Foster C, Zhai Y, Neustadt BR, Stamford AW, Parker EM, Reggiani A, Hunter JC (2009) Characterization of the potent and highly selective 2A receptor antagonists preladenant and SCH 412348 [7-[2-[4-2,4-difluorophenyl]-1-piperazinyl]ethyl]-2-(2-furanyl)-7H-pyrazolo[4,3-e][1,2,4]triazolo[1,5-c]pyrimidin-5-amine] in rodent models of movement disorders and depression. J Pharmacol Exp Ther 330:294–303
71. Thomson Reuters Pharma. https://www.thomson-pharma.com/portal/page/portal/popups/NEWS%20ARTICLE%20DETAILS?_dummy = y&referenceid = 1427557&dbsource = Iddb
72. Shah U, Lankin CM, Boyle CD, Chackalamannil S, Greenlee WJ, Neustadt BR, Cohen-Williams ME, Higgins GA, Ng K, Varty GB, Zhang H, Lachowicz JE (2008) Design, synthesis, and evaluation of fused heterocyclic analogs of SCH 58261 as adenosine A2A receptor antagonists. Bioorg Med Chem Lett 18:4204–4209
73. Neustadt BR, Boyle CD, Chackalamannil S, Harris J, Lankin CM, Liu H, Shah U, Stanford A (2007) 2-Heteroaryl-pyrazolo-[4,3-*e*]-1,2,4-triazolo-[1,5-*c*]-pyrimidine as adenosine A_{2A} receptor antagonists. US-20070066620
74. Neustadt BR, Hao J, Liu H, Boyle CD, Chackalamannil S, Shah UG, Stamford A, Harris JM (2005) Preparation of pyrazolo-[4,3-*e*]-1,2,4-triazolo-[1,5-*c*]-pyrimidine adenosine A_{2A} receptor antagonists. US-20050239795
75. Neustadt BR, Hao J, Liu H, Boyle C, Chackalamannil S, Shah U, Stamford A (2004) 2-Alkynyl- and 2-alkenyl-pyrazolo-[4,3-*e*]-1,2,4-triazolo-[1,5-*c*]-pyrimidine adenosine A_{2A} receptor antagonists. WO-2004094431
76. Shah U, Boyle C, Chackalamannil S, Lankin C, Neustadt B, Stamford A, Cohen-Williams M, Hodgson R, Lachowicz J, Varty G, Zhai Y et al (2010) Identification of structurally distinct analogs of SCH 420814 (preladenant) as adenosine A_{2A} receptor antagonists. 239th American Chemical Society National Meeting, San Francisco, CA, USA: MEDI 547
77. Silverman LS, Caldwell JP, Greenlee WJ, Kiselgof E, Matasi JJ, Tulshian DB, Arik L, Foster C, Bertorelli R, Monopoli A, Ongini E (2007) 3H-[1,2,4]-Triazolo[5,1-i]purin-5-amine derivatives as adenosine A2A antagonists. Bioorg Med Chem Lett 17:1659–1662
78. Harris JM, Neustadt BR, Zhang H, Lachowicz J, Cohen-Williams M, Varty G, Hao J, Stamford AW (2011) Potent and selective adenosine A2A receptor antagonists: [1,2,4]-triazolo[4,3-c]pyrimidin-3-ones. Bioorg Med Chem Lett 21:2497–2501
79. Colotta V, Catarzi D, Varano F, Cecchi L, Filacchioni G, Martini C, Trincavelli L, Lucacchini A (2000) 1,2,4-Triazolo[4,3-*a*]quinoxalin-1-one: a versatile tool for the synthesis of potent and selective adenosine receptor antagonists. J Med Chem 43:1158–1164
80. Colotta V, Catarzi D, Varano F, Filacchioni G, Martini C, Trincavelli L, Lucacchini A (2003) Synthesis of 4-amino-6-(hetero)arylalkylamino-1,2,4-triazolo[4,3-*a*]quinoxalin-1-one derivatives as potent A_{2A} receptor antagonists. Bioorg Med Chem 11:5509–5518
81. Mishra CB, Barodia SK, Prakash A, Senthil Kumar JB, Luthra PM (2010) Novel 8-(furan-2-yl)-3-substituted thiazolo [5,4-e][1,2,4] triazolo[1,5-c] pyrimidine-2(3H)-thione derivatives as potential adenosine A2A receptor antagonists. Bioorg Med Chem 18:2491–2500
82. Shook BC, Rassnick S, Hall D, Rupert KC, Heintzelman GR, Hansen K, Chakravarty D, Bullington JL, Scannevin RH, Magliaro B, Westover L, Carroll K, Lampron L, Russell R, Branum S, Wells K, Damon S, Youells S, Li X, Osbourne M, Demarest K, Tang Y, Rhodes K,

Jackson PF (2010) Methylene amine substituted arylindenopyrimidines as potent adenosine A2A/A1 antagonists. Bioorg Med Chem Lett 20:2864–2867
83. Shook BC, Rassnick S, Osborne MC, Davis S, Westover L, Boulet J, Hall D, Rupert KC, Heintzelman GR, Hansen K, Chakravarty D, Bullington JL, Russell R, Branum S, Wells KM, Damon S, Youells S, Li X, Beauchamp DA, Palmer D, Reyes M, Demarest K, Tang Y-T, Rhodes K, Jackson PF (2010) In vivo characterization of a dual adenosine A2A/A1 receptor antagonist in animal models of Parkinson's disease. J Med Chem 53:8104–8115
84. Shook BC, Rassnick S, Chakravarty D, Wallace N, Ault M, Crooke J, Barbay JK, Wang A, Leonard K, Powell MT, Alford V, Hall D, Rupert KC, Heintzelman GR, Hansen K, Bullington JL, Scannevin RH, Carroll K, Lampron L, Westover L, Russell R, Branum S, Wells K, Damon S, Youells S, Beauchamp D, Li X, Rhodes K, Jackson PF (2010) Optimization of arylindenopyrimidines as potent adenosine A2A/A1 antagonists. Bioorg Med Chem Lett 20:2868–2871
85. Lim H-K, Chen J, Sensenhauser C, Cook K, Preston R, Thomas T, Shook B, Jackson PF, Rassnick S, Rhodes K, Gopaul V, Salter R, Silva J, Evans DC (2011) Overcoming the genotoxicity of a pyrrolidine substituted arylindenopyrimidine as a potent dual adenosine A_{2A}/A_1 antagonist by minimizing bioactivation to an iminium ion reactive intermediate. Chem Res Toxicol 24:1012–1030
86. Shook BC, Rassnick S, Wallace N, Crooke J, Ault M, Chakravarty D, Barbay JK, Wang A, Leonard K, Alford V, Powell MT, Scannevin RH, Carroll K, Lampron L, Westover L, Russell R, Branum S, Wells K, Damon S, Youells S, Li X, Beauchamp DA, Rhodes K, Jackson PF (2012) Design and characterization of optimized adenosine A_{2A}/A_1 receptor antagonists for the treatment of Parkinson's disease. J Med Chem 55:1402–1417
87. Poucher SM, Keddie JR, Singh P, Stoggall SM, Caulkett PW, Jones G, Coll MG (1995) The in vitro pharmacology of ZM 241385, a potent, non-xanthine A_{2A} selective adenosine receptor antagonist. Br J Pharmacol 115:1096–1102
88. Jaakola V-P, Griffith M, Hanson MA, Cherezov V, Chien EYT, Lane JR, Ijzerman AP, Stevens RC (2008) The 2.6 angstrom crystal structure of a human A_{2A} adenosine receptor bound to an antagonist. Science 322:1211–1217
89. Vu CB, Shields P, Peng B, Kumaravel G, Jin X, Phadke D, Wang J, Engber T, Ayyub E, Petter RC (2004) Triamino derivatives of triazolotriazine and triazolopyrimidine as adenosine A_{2A} receptor antagonists. Bioorg Med Chem Lett 14:4835–4838
90. Neustadt BR, Liu H, Hao J, Greenlee WJ, Stamford AW, Foster C, Arik L, Lachowicz J, Zhang H, Bertorelli R, Fredduzzi S, Varty G, Cohen-Williams M, Ng K (2009) Potent and selective adenosine A2A receptor antagonists: 1,2,4-triazolo[1,5-c]pyrimidines. Bioorg Med Chem Lett 19:967–971
91. Matasi JJ, Caldwell JP, Zhang H, Fawzi A, Higgins GA, Cohen-Williams ME, Varty GB, Tulshian DB (2005) 2-(2-Furanyl)-7-phenyl[1,2,4]triazolo[1,5-c]pyrimidin-5-amine analogs as adenosine A2A antagonists: the successful reduction of hERG activity. Bioorg Med Chem Lett 15:3675–3678
92. Peng H, Kumaravel G, Yao G, Sha L, Wang J, Van Vlijmen H, Bohnert T, Huang C, Vu CB, Ensinger CL, Chang H, Engber TM, Whalley ET, Petter RC (2004) Novel bicyclic piperazine derivatives of triazolotriazine and triazolopyrimidines as highly potent and selective adenosine A_{2A} receptor antagonists. J Med Chem 47:6218–6229
93. Dowling JE, Vessels JT, Haque S, Chang HX, van Vloten K, Kumaravel G, Engber T, Jin X, Phadke D, Wang J, Ayyub E, Petter RC (2005) Synthesis of [1,2,4]triazolo[1,5-*a*]pyrazines as adenosine A_{2A} receptor antagonists. Bioorg Med Chem Lett 15:4809–4813
94. Yao G, Haque S, Sha L, Kumaravel G, Wang J, Engber TM, Whalley ET, Onlon PR, Chang H, Kiesman WF, Petter RC (2005) Synthesis of alkyne derivatives of a novel triazolopyrazine as A_{2A} adenosine receptor antagonists. Bioorg Med Chem Lett 15:511–515
95. Holschbach MH, Bier D, Stusgen S, Wutz W, Sihver W, Coenen HH, Olsson RA (2006) Synthesis and evaluation of 7-amino-2-(2(3)-furyl)-5-phenethylamino-oxazolo[5,4-*d*]

pyrimidines as potential A_{2A} adenosine receptor antagonists for positron emission tomography (PET). Eur J Med Chem 41:7–15
96. Shao Y, Cole AG, Brescia M-R, Qin L-Y, Duo J, Stauffer TM, Rokosz LL, McGuinness BF, Henderson I (2009) Synthesis and SAR studies of trisubstituted purinones as potent and selective adenosine A2A receptor antagonists. Bioorg Med Chem Lett 19:1399–1402
97. Weiss SM, Benwell K, Cliffe IA, Gillespie RJ, Knight AR, Lepiniere J, Misra A, Pratt RM, Revell D, Upton R, Dourish CT (2003) Discovery of nonxanthine adenosine A_{2A} receptor antagonists for the treatment of Parkinson's disease. Neurology 61:S101–S106
98. Gillespie RJ, Adams DR, Bebbington D, Benwell K, Cliffe IA, Dawson CE, Dourish CT, Fletcher A, Gaur S, Giles PR, Jordan AM, Knight AR, Knutsen LJS, Lawrence A, Lerpiniere J, Misra A, Porter RHP, Pratt RM, Shepherd R, Upton R, Ward SE, Weiss SM, Williamson DS (2008) Antagonists of the human adenosine A_{2A} receptor. Part 1: discovery and synthesis of thieno[3,2-*d*]pyrimidine-4-methanone derivatives. Bioorg Med Chem Lett 18:2916–2919
99. Gillespie RJ, Cliffe IA, Sawson CE, Dourish CT, Gaur S, Giles PR, Jordan AM, Knight AR, Lawrence A, Lerpiniere J, Misra A, Pratt RM, Todd RS, Upton R, Weiss SM, Williamson DS (2008) Antagonists of the human adenosine A_{2A} receptor. Part 2: design and synthesis of 4-arylthieno[3,2-*d*]pyrimidine derivatives. Bioorg Med Chem Lett 18:2920–2923
100. Gillespie RJ, Cliffe IA, Dawson CE, Dourish CT, Gaur S, Jordan AM, Knight AR, Lerpiniere J, Misra A, Pratt RM, Roffey J, Stratton GC, Upton R, Weiss SM, Williamson DS (2008) Antagonists of the human adenosine A2A receptor. Part 3: design and synthesis of pyrazolo[3,4-d]pyrimidines, pyrrolo[2,3-d]pyrimidines and 6-arylpurines. Bioorg Med Chem Lett 18:2924–2929
101. Gillespie RJ, Bamford SJ, Botting R, Comer M, Denny S, Gaur S, Griffin M, Jordan AM, Knight AR, Lerpiniere J, Leonardi S, Lightowler S, McAteer S, Merrett A, Misra A, Padfield A, Reece M, Saadi M, Selwood DL, Stratton GC, Surry D, Todd R, Tong X, Ruston V, Upton R, Weiss SM (2009) Antagonists of the human A2A adenosine receptor. 4. Design, synthesis, and preclinical evaluation of 7-aryltriazolo[4,5-d]pyrimidines. J Med Chem 52:33–47
102. Vernalis (2004) Vernalis and Biogen Idec to collaborate on research for Parkinson's disease. Press Release, June 24
103. Vernalis (2010) Vernalis announces A_{2A} receptor antagonist programme for Parkinson's disease with next generation compound. Press Release, July 16
104. Vernalis (2012) Positive results achieved in Vernalis' receptor occupancy study of V81444 for Parkinson's disease and other CNS indications. Press Release, December 12
105. Minetti P, Tinti MO, Carminati P, Castorina M, Di Cesare MA, Di Serio S, Gallo G, Ghirardi O, Giorgi F, Giorgi L, Piersanti G, Bartoccini F, Tarzia G (2005) 2-n-Butyl-9-methyl-8-[1,2,3]triazol-2-yl-9H-purin-6-ylamine and analogues as A2A adenosine receptor antagonists. Design, synthesis, and pharmacological characterization. J Med Chem 48:6887–6896
106. Stasi MA, Borsini F, Varani K, Vincenzi F, Di Cesare MA, Minetti P, Ghirardi O, Carminati P (2006) ST 1535: a preferential A2A adenosine receptor antagonist. Int J Neuropsychopharmacol 9:575–584
107. Rose S, Ramsay Croft N, Jenner P (2007) The novel adenosine A2a antagonist ST1535 potentiates the effects of a threshold dose of L-dopa in unilaterally 6-OHDA-lesioned rats. Brain Res 1133:110–114
108. Tronci E, Simola N, Borsini F, Schintu N, Frau L, Carminati P, Morelli M (2007) Characterization of the antiparkinsonian effects of the new adenosine A2A receptor antagonist ST1535: acute and subchronic studies in rats. Eur J Pharmacol 566:94–102
109. Rose S, Jackson MJ, Smith LA, Stockwell K, Johnson L, Carminati P, Jenner P (2006) The novel adenosine A2a receptor antagonist ST1535 potentiates the effects of a threshold dose of L-DOPA in MPTP treated common marmosets. Eur J Pharmacol 546:82–87

110. Cabri W, Minetti P, Piersanti G, Tarsia G (2010) Oxidated derivatives of triazolylpurines useful as ligands of the adenosine A_{2A} receptor and their use as medicaments. PCT Int. Appl. WO 2010106145
111. Shook BC, Chakravarty D, Barbay JK, Wang W, Leonard K, Alford V, Powell M, Beauchamp DA, Rassnick S, Scannevin R, Carroll K, Wallace N, Crooke J, Ault M, Lampron L, Westover L, Rhodes K, Jackson PF (2011) Aminomethyl substituted thieno [2,3-d]pyrimidines as adenosine A2A receptor antagonists. Med Chem Commun. doi:10.1039/C1MD00082A
112. Shook BC, Chakravarty D, Barbay JK, Wang A, Leonard K, Alford V, Powell MT, Rassnick S, Scannevin RH, Carroll K, Wallace N, Crooke J, Ault M, Lampron L, Westover L, Rhodes K, Jackson PF (2013) Substituted thiean[2,3-*d*]pyrimidines as adenosine A_{2A} receptor antagonists. Bioorg Med Chem Lett 23:2688–2691
113. McGuinness BF, Cole AG, Dong G, Brescia M-R, Shao Y, Henderson I, Rokosz LL, Stauffer TM, Mannava N, Kimble EF, Hicks C, White N, Wines PG, Quadros E (2010) Discovery of 2-aminoimidazopyridine adenosine A2A receptor antagonists. Bioorg Med Chem Lett 20:6845–6849
114. Flohr A, Moreau JL, Poli SM, Riemer C, Steward L (2005) 4-Hydroxy-4-methyl-piperidine-1-carboxylic acid (4-methoxy-7-morpholin-4-yl-benzothiazol-2-yl)-amide. US 20050261289
115. Flohr A, Riemer C (2006) Benzothiazole derivatives. WO 2006008041
116. Flohr A, Riemer C (2006) Substituted benzothiazoles. WO 2006008040
117. Norcross RD (2004) Benzoxazole derivatives and their use as adenosine receptor ligands. PCT int. Appl. WO 2004063177
118. Norcross RD (2005) Thiazolopyridine. U.S. Pat. Appl. Publ. US 20050261289
119. Woiwode T, Noran M (2000) 4-Hydroxy-4-methyl-piperidine-1-carboxylic acid (4-methoxy-7-morpholin-4-yl-benzothiazol-2-yl)-amide for the treatment of post-traumatic stress disorder. PCT Int. Appl. WO 2009015236
120. Therapeutics S (2007) Roche and Synosis Therapeutics announce partnership to explore potential of five compounds targeting the central nervous system. Press Release, January, 4
121. Therapies B (2012) Biotie's tozadenant (SYN115) meets primary and multiple secondary endpoints in phase 2b study in Parkinson's disease. Press Release, December, 11
122. Mayer S, Schann S (2010) New adenosine receptor ligands and uses thereof. PCT Int. Appl. WO 2010084425
123. Shiohara H, Nakamura T, Kobayashi S (2006) Novel benzofuran derivative, pharmaceutical composition comprising the same, and use of the derivative or composition. PCT Int. Appl. WO 2006115134
124. Saku O, Saki M, Kurokawa M, Ikeda K, Takizawa T, Uesaka N (2010) Synthetic studies on selective adenosine A2A receptor antagonists: synthesis and structure-activity relationships of novel benzofuran derivatives. Bioorg Med Chem Lett 20:1090–1093
125. Saku O, Saki M, Kurokawa M, Ikeda K, Uchida S-I, Takizawa T, Uesaka N (2010) Synthetic studies on selective adenosine A2A receptor antagonists. Part II: synthesis and structure-activity relationships of novel benzofuran derivatives. Bioorg Med Chem Lett 20:3768–3771
126. Yonishi S, Aoki S, Matsushima Y, Akahane A (2005) Preparation of pyrazines as adenosine A_1 and A_{2A} receptor antagonists and their pharmaceutical compositions. PCT Int. Appl. WO 2005040151
127. Mihara T, Mihara K, Yarimizu J, Mitani Y, Matsuda R, Yamamoto H, Aoki S, Akahane A, Iwashita A, Matsuoka N (2007) Pharmacological characterization of a novel, potent adenosine A1 and A2A receptor dual antagonist, 5-[5-amino-3-(4-fluorophenyl)pyrazin-2-yl]-1-isopropylpyridine-2(1H)-one (ASP5854), in models of Parkinson's disease and cognition. J Pharmacol Exp Ther 323:708–719
128. Mihara T, Noda A, Arai H, Mihara K, Iwashita A, Murakami Y, Matsuya T, Miyoshi S, Nishimura S, Matsuoka N (2008) Brain adenosine A2A receptor occupancy by a novel A1/A2A receptor antagonist, ASP5854, in rhesus monkeys: relationship to anticataleptic effect. J Nucl Med 49:1183–1188

129. Mihara T, Iwashita A, Matsuoka N (2008) A novel adenosine A1 and A2A receptor antagonist ASP5854 ameliorates motor impairment in MPTP-treated marmosets: comparison with existing anti-Parkinson's disease drugs. Behav Brain Res 194:152–161
130. Sams AG, Mikkelsen GK, Larsen M, Torup L, Brennum LT, Schroder TJ, Bang-Andersen B (2010) Hit-to-lead optimization of a series of carboxamides of ethyl 2-amino-4-phenylthiazole-5-carboxylates as novel adenosine A2A receptor antagonists. Bioorg Med Chem Lett 20:5241–5244
131. Cole AG, Stauffer TM, Rokosz LL, Metzger A, Dillard LW, Zeng W, Henderson I (2009) Synthesis of 2-amino-5-benzoyl-4-(2-furyl)thiazoles as adenosine A2A receptor antagonists. Bioorg Med Chem Lett 19:378–381
132. Langmead CJ, Andrews SP, Congreve M, Errey JC, Hurrell E, Marshall FH, Mason JS, Richardson CM, Robertson N, Zhukov A, Weir M (2012) Identification of novel adenosine A_{2A} receptor antagonists by virtual screening. J Med Chem 55:1904–1909
133. Congreve MS, Andrews SP, Mason JS, Richardson CM, Brown GA (2011) 1,2,4-triazine-4-amine derivatives. PCT Int. Appl. WO 2011095625
134. Congreve M, Andrews SP, Dore AS, Hollenstein K, Hurrell E, Langmead CJ, Mason JS, Ng IW, Tehan B, Zhukov A, Weir M, Marshall FH (2012) Discovery of 1,2,4-triazine derivatives as adenosine A_{2A} antagonists using structure based drug design. J Med Chem 55:1898–1903
135. Camacho Gomez JA, Castro-Palomino Laria JC (2011) 4-Aminopyrimidine derivatives and their use as adenosine A_{2A} receptor antagonists. PCT Int. Appl. WO 2011121418
136. Zhang X, Rueter JK, Chen Y, Moorjani M, Lanier MC, Lin E, Gross RS, Tellew JE, Williams JP, Lechner SM, Markison S, Joswig T, Malany S, Santos M, Castro-Palomino JC, Crespo MI, Prat M, Gual S, Diaz J-L, Saunders J, Slee DH (2008) Synthesis of N-pyrimidinyl-2-phenoxyacetamides as adenosine A2A receptor antagonists. Bioorg Med Chem Lett 18:1778–1783
137. Zhang X, Tellew JE, Luo Z, Moorjani M, Lin E, Lanier MC, Chen Y, Williams JP, Saunders J, Lechner SM, Markison S, Joswig T, Petroski R, Piercey J, Kargo W, Malany S, Santos M, Gross RS, Wen J, Jalalil K, O'Brien Z, Stotz CE, Crespo MI, Diaz J-L, Slee DH (2008) Lead optimization of 4-acetylamino-2-(3,5-dimethylpyrazol-1-yl)-6-pyridylpyrimidines as A_{2A} adenosine receptor antagonists for the treatment of Parkinson's disease. J Med Chem 51:7099–7110
138. Lanier MC, Moorjani M, Luo Z, Chen Y, Lin E, Tellew JE, Zhang X, Williams JP, Gross RS, Lechner SM, Markison S, Joswig T, Kargo W, Piercey J, Santos M, Malany S, Zhao M, Petroski R, Crespo MI, Diaz J-L, Saunders J, Wen J, O'Brien Z, Jalali K, Madan A, Slee DH (2009) *N*-[6-Amino-2-(heteroaryl)-pyrimidine-4-yl]acetamides as A_{2A} receptor antagonists with improved druglike properties and in vivo efficacy. J Med Chem 52:709–717
139. Slee DH, Zhang X, Moorjani M, Lin E, Lanier MC, Chen Y, Rueter JK, Lechner SM, Markison S, Malany S, Joswig T, Santos M, Gross RS, Williams JP, Castro-Palomino JC, Crespo MI, Prat M, Gual S, Diaz J-L, Wen J, O'Brien Z, Saunders J (2008) Identification of novel, water-soluble, 2-amino-N-pyrimidin-4-yl acetamides as A2A receptor antagonists with in vivo efficacy. J Med Chem 51:400–406
140. Slee DH, Chen Y, Zhang X, Moorjani M, Lanier MC, Lin E, Rueter JK, Williams JP, Lechner SM, Markison S, Malany S, Santos M, Gross RS, Jalali K, Sai Y, Zuo Z, Yang C, Castro-Palomino JC, Crespo MI, Prat M, Gual S, Diaz J-L, Saunders J (2008) 2-Amino-N-pyrimidin-4-ylacetamides as A2A receptor antagonists: 1. Structure-activity relationships and optimization of heterocyclic substituents. J Med Chem 51:1719–1729
141. Moorjani M, Luo Z, Lin E, Vong BG, Chen Y, Zhang X, Rueter JK, Gross RS, Lanier MC, Tellew JE, Williams JP, Lechner SM, Malany S, Santos M, Crespo MI, Diaz J-L, Saunders J, Slee DH (2008) 2,6-Diaryl-4-acylaminopyrimidines as potent and selective adenosine A2A antagonists with improved solubility and metabolic stability. Bioorg Med Chem Lett 18:5402–5405

142. Gillespie RJ, Bamford SJ, Gaur S, Jordan AM, Lerpiniere J, Mansell HL, Stratton GC (2009) Antagonists of the human A2A receptor. Part 5: highly bio-available pyrimidine-4-carboxamides. Bioorg Med Chem Lett 19:2664–2667
143. Gillespie RJ, Bamford SJ, Clay A, Gaur S, Haymes T, Jackson PS, Jordan AM, Klenke B, Leonardi S, Liu J, Mansell HL, Ng S, Saadi M, Simmonite H, Stratton GC, Todd RS, Williamson DS, Yule IA (2009) Antagonists of the human A2A receptor. Part 6: further optimization of pyrimidine-4-carboxamides. Bioorg Med Chem 17:6590–6605
144. Mishina M, Ishiwata K, Kimura Y, Naganawa M, Oda K, Kobayashi S, Katayama Y, Ishii K (2007) Evaluation of distribution of adenosine A_{2A} receptors in normal human brain measured with [^{11}C]-TMSX PET. Synapse 61:778–784
145. Mishina M, Ishiwata K, Kimura Y, Naganawa M, Kitamura S, Suzuki M, Hashimoto M, Ishibashi K, Oda K, Sakata M, Hamamoto M, Kobayashi S, Katayama Y, Ishii K (2011) Adenosine A_{2A} receptors measured with [^{11}C]-TMSX PET in the striata of Parkinson's disease patients. PLoS One 6:e17338
146. Zocchi C, Ongini E, Ferrara S, Baraldi PG, Dionisotti S (1996) Binding of the radioligand [^{3}H]-SCH58261, a new non-xanthine A_{2A} adenosine receptor antagonist, to rat striatal membranes. Br J Pharmacol 117:1381–1386
147. Belardinelli L, Shryok JC, Ruble J, Monopoli A, Dionisotti S, Ongini E, Dennis DM, Baker SP (1996) Binding of the novel nonxanthine A_{2A} adenosine receptor antagonist [^{3}H]-SCH58261 to coronary artery membranes. Circ Res 79:1153–1160
148. Dionisotti S, Ferrara S, Molta C, Zocchi C, Ongini E (1996) Labeling of A_{2A} adenosine receptors in human platelets by use of the new nonxanthine antagonist radioligand [^{3}H]-SCH58261. J Pharmacol Exp Ther 278:1209–1214
149. Varani K, Gessi S, Dionisotti S, Ongini E, Borea PA (1998) [^{3}H]-SCH58261 labeling of functional A_{2A} adenosine receptors in human neutrophil membranes. Br J Pharmacol 123:1723–1731
150. Fredholm BB, Lindstrom K, Dionisotti S, Ongini E (1998) [^{3}H]-SCH58261, a selective adenosine A_{2A} receptor antagonist, is a useful ligand in autoradiographic studies. J Neurochem 70:1210–1216
151. Todde S, Moresco RM, Simonelli P, Baraldi PG, Cacciari B, Spalluto G, Varani K, Monopoli A, Matarrese M, Carpinelli A, Magni F, Kienle MG, Fazio F (2000) Design, radiosynthesis, and biodistribution of a new potent and selective ligand for in vivo imaging of the adenosine A_{2A} receptor system using positron emission tomography. J Med Chem 43:4359–4362
152. Brooks DJ, Papapetropoulos S, Vandenhende F, Tomic D, Coppell A, O'Neill G (2010) An open-label, positron emission tomography study to assess adenosine A_{2A} brain receptor occupancy of vipadenant (BIIB014) at steady-state levels in healthy male volunteers. Clin Neuropharmacol 33:55–60
153. Niswender CM, Johnson KA, Weaver CD, Jones CK, Xiang Z, Luo Q, Rodriguez AL, Marlo JE, de Paulis T, Thompson AD, Days EL, Nalywajko T, Austin CA, Baxter Williams M, Ayala JE, Williams R, Lindsley CW, Conn PJ (2008) Discovery, characterization, and antiparkinsonian effect of novel positive allosteric modulators of metabotropic glutamate receptor 4. Mol Pharmacol 74:1345–1358

Top Med Chem (2016) 18: 43–110
DOI: 10.1007/7355_2014_66
© Springer-Verlag Berlin Heidelberg 2014
Published online: 2 December 2014

Targeting α-Synuclein as a Parkinson's Disease Therapeutic

Luke A. Esposito

Abstract α-Synuclein is a dynamic protein capable of assuming an ensemble of physiological and pathological structures. Its primary role in Parkinson's disease (PD) pathogenesis makes it an attractive though nonconventional therapeutic target. An understanding of α-synuclein intra- and intermolecular interactions and post-translational modifications that lead to its misfolding, aggregation, accumulation, and cell-to-cell prion-like spreading is necessary to inform the design of α-synuclein-directed therapeutics. Native α-synuclein interacts with membranes and plays an important role in synaptic transmission and vesicle trafficking at the presynaptic terminal, though aggregated α-synuclein may be compromised in its ability to perform these functions. In addition to discussing α-synuclein structural biology, this chapter explores the variety of experimental therapeutic approaches currently under investigation that aim to maintain physiological α-synuclein levels, limit toxic aggregates, and lessen effects of misfolded α-synuclein on cellular homeostasis. For example, oligonucleotide-based approaches limit α-synuclein gene expression. In addition, select small molecules and peptides inhibit α-synuclein aggregation at substoichiometric concentrations in favor of less structured, more soluble, and less toxic oligomers that may be better substrates for clearance. Some small molecules also remodel existing α-synuclein fibrils. Modest chemical changes to these small molecules can greatly impact their mechanism of action. Finally, α-synuclein-directed immunotherapy enhances the lysosomal clearance of α-synuclein in experimental models and is currently in clinical trials. Other therapeutic approaches target α-synuclein posttranslational modifications, aim to limit its impact on mitochondria or its ability to alter gene expression in the nucleus.

Keywords Amyloid, Autophagy, Disease-modifying, Immunotherapy, Lewy body, Natively unfolded, Nutraceutical, Off-pathway oligomers, Parkinson's disease, Protein misfolding, Reactive oxygen species, Small molecule, α-Synuclein

L.A. Esposito (✉)
ProteoTech Inc., 12040 115th Ave, NE, Kirkland, WA 98034, USA
e-mail: esposito@proteotech.com

Contents

1 Introduction

Novel approaches to treat PD, like those directed at many of the other eponymous neurodegenerative disorders characterized by pathogenic misfolded proteins, have rapidly evolved in recent years. Research and drug discovery efforts – initially aimed at understanding and treating the anatomical and neurochemical changes that cause the clinical syndrome – have expanded rapidly as a result of the full biochemical characterization of the pathological lesion and underlying molecular culprits. Hence, in the last 15 years, discoveries in molecular genetics have been a critical driving force behind the search for new medicines that target underlying disease etiology – so-called disease-modifying therapeutics.

It was within this new framework that intracellular lesions known as Lewy bodies first revealed the central role of α-synuclein in the pathogenesis and progression [1, 2] of PD and related "synucleinopathies" such as dementia with Lewy bodies (DLB) and multiple systems atrophy (MSA) (Fig. 1). Lewy bodies consist

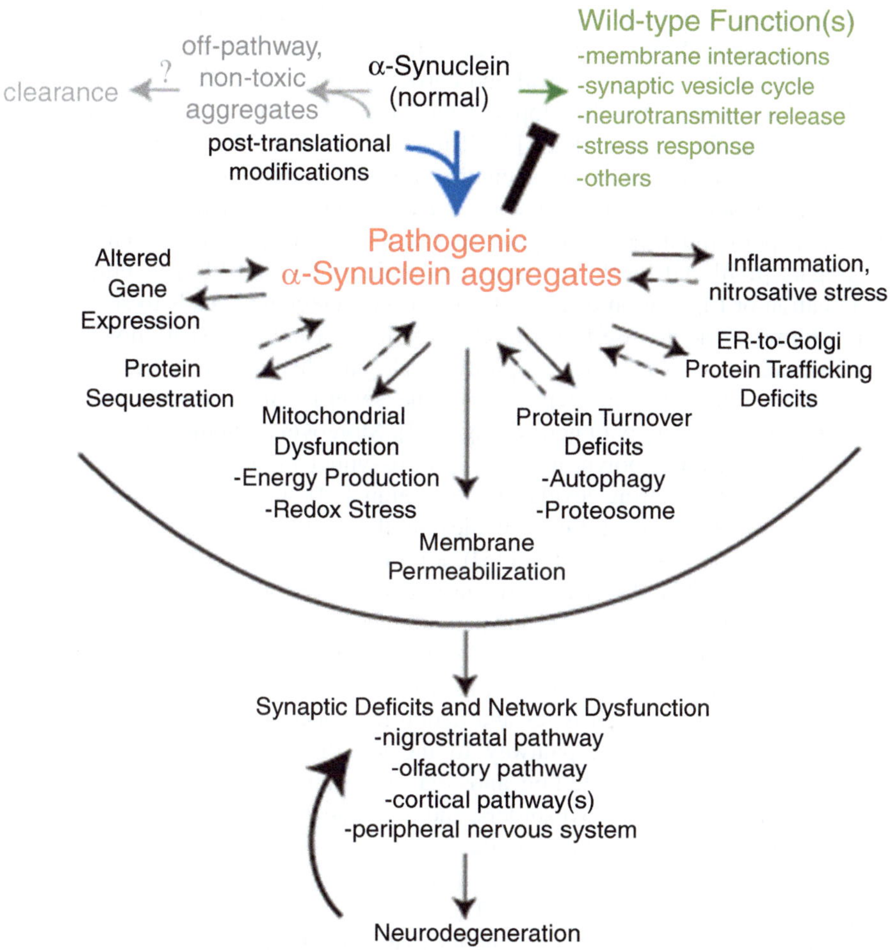

Fig. 1 Pathogenic mechanisms of aggregated α-synuclein. In its normal state, α-synuclein is a natively unfolded protein, helical tetramer, or another conformer whose wild-type functions (indicated by *green*) are not fully defined but likely involve a role in synaptic vesicle cycling, neurotransmitter release at the synapse, and stress response. Many of these wild-type functions are mediated by the ability of α-synuclein to bind membranes and populate a variety of structural conformations. However, when relatively rare misfolding events (*blue arrow*) occur under specific cellular conditions or as a result of posttranslational modifications (*blue curved line*) or genetic mutations (not shown), pathogenic aggregates form (indicated in *red*). These heterogeneous aggregates likely include toxic oligomeric species and β-sheet-rich fibrils that can disrupt α-synuclein wild-type function and/or confer toxic gain of function (*black arrows*). Moreover, most of the deleterious changes also increase steady-state levels of α-synuclein aggregates (*dashed lines* with *small arrowheads*), for example, by limiting their turnover or creating conditions (i.e., nitrosative stress, redox stress, protein sequestration, or decreased energy production) that promote their formation. Many of the deleterious effects of α-synuclein aggregates are interrelated, but the illustration of such relationships is omitted for simplicity. For example, α-synuclein aggregates can sequester transcription factors in the cytosol, resulting in decreased autophagic clearance. α-Synuclein-induced synaptic deficits and network dysfunction result from a collection of events

primarily of an eosinophilic dense core surrounded by a halo of radiating filamentous α-synuclein fibrils. Lewy bodies are reliably detected in the postmortem brains of PD patients in both subcortical and cortical regions, including the nigrostriatal pathway, the cortex, and the hippocampus [1, 3–5], with the extent of pathology depending on the disease stage [2]. The critical role of the small but dynamic α-synuclein was corroborated soon thereafter by extensive molecular genetic, biochemical, and cell biological studies. These studies revealed putative "gain-of-function" point mutations in α-synuclein that increase its propensity to misfold and aggregate [6–8] and result in autosomal-dominant familial PD [9]. Furthermore, duplication or triplication of the wild-type α-synuclein gene leads to severe, early onset forms of the disease [10–12], characterized by both classical motor features (bradykinesia, resting tremor, rigidity, and gait disturbances) and non-motor impairments, including dementia, with the severity of symptoms and rate of progression increasing with gene dosage [13]. Moreover, genome-wide association studies have identified polymorphisms at the α-synuclein locus that increase disease risk, possibly by altering levels of the protein [14]. It is worth noting the rare exception of PD lacking Lewy bodies in the substantia nigra resulting from mutations in parkin (reviewed in [15]), a protein with a direct role in Lewy body formation [16]. This observation suggests that PD is not one disease but rather a syndrome of multiple etiologies and pathological manifestations. Nevertheless, the overwhelming majority of PD cases, both idiopathic and inherited, exhibit extensive α-synuclein pathology. Indeed, misfolded α-synuclein exerts its deleterious effects in a number of cell types and neural networks, extending beyond just the dopaminergic nigrostriatal pathway to include the peripheral nervous system. Non-motor symptoms, including anosmia, gastrointestinal tract disorders, rapid eye movement (REM) sleep disorders, and psychiatric disorders such as depression are increasingly recognized among the earliest PD symptoms and critical determinants of quality of life for patients, in part because some are poorly responsive to dopaminergic therapies [17, 18]. Therefore, a therapeutic aimed at limiting α-synuclein toxicity may have benefits that extend beyond those gained from therapies that target specific neurotransmitter systems. Thus, a theme of this review is novel therapeutics that limit steady-state levels of specific toxic α-synuclein forms or block its downstream toxic effects.

Fig. 1 (continued) that over time lead to neurodegeneration in select brain regions, leading to clinical PD. Therefore, blocking the formation of pathogenic aggregates (*blue arrow*) or any of their downstream effectors (*black arrows* in top half of the figure) is the prime target for disease-modifying interventions. Alternatively, α-synuclein-directed therapeutics redirect α-synuclein to off-pathway, nontoxic aggregates (*gray arrow*) that may be good substrates for clearance

2 α-Synuclein Structure, Function, and Pathogenic Misfolding

2.1 α-Synuclein Populates an Ensemble of Native and Pathogenic Structures

α-Synuclein consists of 140 residues, lacking both Cys and Trp residues, and with a low predicted overall hydrophobicity. Its primary structural motifs consist of an N-terminal lipid-binding region (residues 1–60) capable of assuming helical character and containing imperfect 11-residue amphipathic repeats with the KTGEGV consensus sequence, a central hydrophobic region (residues 61–95) known as NAC (or non-amyloid component, originally discovered at autopsy in the brains of Alzheimer's disease patients) believed to be critical for amyloid formation, and a highly acidic C-terminal region (residues 96–140) that favors extended structures. As a result of the highly acidic C-terminal region, at neutral pH, α-synuclein has a large net negative charge of approximately –9 [19].

Unlike so many classic targets of drug discovery, including receptors, transporters, and ion channels, α-synuclein populates a wide variety of structures and conformations. In addition, its wild-type function is not fully understood, although it likely involves vesicle trafficking and modulation of dopamine release at the synapse. While considered by many to be a predominately natively unfolded monomeric protein [20–22], it has been recently asserted that α-synuclein in its native form is a helical tetramer [23–25] or that it populates a set of metastable multimeric conformers [26] distinct from the misfolded putatively toxic oligomers described in Sect. 2.5. In any case, it is generally accepted that the amyloidogenic precursor is a monomer (possibly a disassembled multimer or the true native form), and this free monomeric state of α-synuclein can populate a highly heterogeneous ensemble of natively unfolded or partially folded structures that may or may not coexist in vivo. α-Synuclein polymerizes through a nucleation-dependent process [27] and likely populates partially folded [20], β-like [28], monomeric, and metastable multimeric (oligomeric) states [29, 30], a subset of which seed (kinetically accelerate) further addition of monomer on the way to the thermodynamically stable, fibrillar form that resembles classic amyloid [31]. A corollary to this hypothesis is that a structural conversion from the initially formed unstructured oligomers to more stable, β-sheet-rich toxic oligomers is the key step in fibril formation [32]. Oligomeric forms of α-synuclein also include conformations that are incapable of nucleating or promoting further aggregation. These "off-pathway" oligomers are thus not precursors to large β-sheet-rich fibrils and are less likely to dissociate to monomers than transient oligomers but may instead coexist with fibrils in the steady-state phase [29]. For many, if not most of these structures, from monomer to oligomer to fibril, a dynamic equilibrium likely exists, the balance of which can depend on a number of conditions (discussed in greater detail below) and ultimately determine whether α-synuclein behaves normally or assumes an ectopic deleterious structure/function.

More evidence of its chameleonic and multifaceted nature is found at the cell biological level. For example, α-synuclein is abundant in both brain and blood. It is typically found in both the cytosolic (soluble) and membrane-associated (particulate) fractions of the brain, and it may be present at both the synapse and the nucleus, as reflected in its name [33]. However, it localizes primarily to the presynaptic nerve terminal under normal conditions, and its pathogenic forms may be transmitted from cell to cell along neural networks (reviewed in [34]). α-Synuclein pathology may even originate outside the central nervous system in the enteric nervous system [35], consistent with constipation being one of the earliest PD symptoms, preceding motor dysfunction by many years [36]. With such a wide cellular and subcellular distribution, and spectrum of possible structures with potentially overlapping but separate pathogenic mechanisms, where does one start when attempting to target α-synuclein as a PD therapeutic?

Intensive research and drug discovery efforts are currently underway with the primary goal of targeting α-synuclein using a variety of approaches including those aimed at eliminating or lessening the abundance of the pathogenic forms directly and lessening their downstream deleterious effects. While great challenges exist in targeting a natively unfolded and dynamic target with a wild-type function, identification of a disease-modifying therapeutic will likely focus, at least in part, on the misfolded forms that are critical molecular determinants of disease etiology and progression.

2.2 Natively "Unfolded" Monomers Are the Amyloidogenic Precursor

Whether derived from a disassembled multimeric conformer or existing in its native state, monomeric α-synuclein is believed to be the amyloidogenic precursor, a critical part of the nucleation-dependent polymerization that gives rise to potentially toxic intermediates and amyloid fibrils. The monomeric form may also exert its wild-type function (see Sect. 2.4), thus providing further rationale for therapeutic approaches that favor monomeric α-synuclein.

2.3 Intramolecular Interactions as Determinants of Aggregation

Prior to its integration into amyloid, and its full adoption of structures that lead to amyloid formation, the monomeric state of α-synuclein populates a highly heterogeneous ensemble of unfolded or partially folded structures that may or may not coexist in vivo.

As revealed by a variety of spectroscopic studies, including solution-state nuclear magnetic resonance (NMR), monomeric α-synuclein can be considered a natively unfolded protein, although it is more compact, on average, than pure random coil [37]. Molecular modeling experiments also predict structural characteristics of an unfolded protein, including greater surface area exposed to solvent and a less compact structure, comprised of conformations with greater radii of gyration, indicative of more intramolecular movement, relative to more compact globular states [38]. Why might an abundant and disease-relevant protein evolve to exist in a natively unfolded state, with the consequential increased propensity for aggregation? A natively unfolded state may provide functional benefits in the complex and crowded cellular milieu, for example, by creating greater surface area for binding and a more "flexible" functionality. These characteristics may, in turn, allow adaption to cellular conditions such as folding upon binding to the target (i.e., membranes in the case of α-synuclein [39, 40]) or service as a linker in the assembly of macromolecular arrays [41].

The amino acid composition of α-synuclein predicts secondary structure motifs. Indeed, even in its "unfolded" state, α-synuclein is not completely devoid of transient and/or more stable intramolecular interactions and secondary structure motifs. For example, the degree of signal broadening demonstrated in paramagnetic relaxation enhancement (PRE) NMR studies confirms that α-synuclein is, on average, more compact than expected for a random coil protein [42]. These studies also suggest the existence of long-range interactions between the N-terminus and C-terminus and between the C-terminus and NAC region. The relevance of such residual secondary structure and intramolecular interactions to the aggregation propensity of α-synuclein was first explored in an attempt to explain the accelerated aggregation rate of A30P mutant relative to wild-type α-synuclein. The A30P mutation [43] is the second of five α-synuclein point mutations that have been identified to date including A53T [9], E46K [44], and the recently discovered G51D and H50Q [45–47], all of which cluster in the N-terminal region of α-synuclein. Using high-resolution NMR, it was demonstrated that the A30P mutation disrupted residual helical content at the N-terminus. In wild-type α-synuclein, this helical region may provide a hydrophobic face that facilitates an interaction with the NAC region, thereby shielding the NAC region from intermolecular contacts. As a helix breaker, the proline substitution mutation relieves this interference and promotes oligomerization. The A53T mutation, however, has a more direct effect on the NAC region, favoring a slightly extended conformation that could facilitate intermolecular interactions [48]. Subsequent work demonstrated interactions between a hydrophobic cluster in the NAC region (amino acids 85–95) and the C-terminus (residues 110–130) and an electrostatic interaction between the C-terminal amino acids centered at residue 120 and the N-terminus, around residue 20 [49, 50]. These interactions could occlude the hydrophobic NAC region [50, 51] and thereby inhibit intermolecular interactions that involve this region, having the net effect of preventing misfolding into structures that give rise to higher-order, potentially more toxic (i.e., oligomeric and fibrillar) forms. The notion of the C-terminus shielding the amyloidogenic NAC region is consistent with α-synuclein C-terminal truncations promoting misfolding [19, 52]. In addition, comparison of human α-synuclein with the more aggregation-prone mouse α-synuclein suggests that differences in intramolecular interactions could account for the different aggregation propensities, with mouse α-synuclein less prone to stabilizing intramolecular interactions relative to human α-synuclein [37].

More recently, however, the significance of intramolecular interactions on α-synuclein misfolding propensity has been challenged, with the assertion that long-range interactions shielding solvent exposure of the NAC region have only a minimal effect on oligomerization [53]. Instead, intermolecular electrostatic repulsion of the negatively charged C-terminus, hydrophobicity, and overall structural compactness may play a more prominent role in oligomerization [54–57]. For example, using single-molecule (sm) fluorescence resonance energy transfer (FRET), it was demonstrated that at low pH, charge neutralization at the C-terminus increases intramolecular contacts and allows inclusion into a hydrophobic cluster that is more aggregation prone [56]. Polycations, on the other hand, may shield the negative charges at the C-terminus, but not eliminate them, without causing significant conformational changes, thereby altering aggregation propensity by reducing electrostatic repulsion between protein molecules without requiring formation of the hydrophobic core found at low pH [56]. Notably, the C-terminus is phosphorylated in vivo at multiple sites [58], and this has been shown in some reports to increase aggregation (discussed in [59] and Sect. 3.3) consistent with the proper balance of electrostatic interactions involving the C-terminus being a determinant of aggregation propensity (although some phosphorylated forms, such as Ser87-P, may inhibit aggregation [60]).

Some of the differences in opinion about the role of intramolecular long-range interactions in α-synuclein aggregation should be viewed in the context of the limitations of the experimental methods used to draw conclusions, including the uncertainty associated with molecular modeling and the potential for structural changes resulting from introduction of a paramagnetic probe into the protein for PRE NMR analysis.

A related hypothesis focuses largely on internal protein dynamics as a determinant of aggregation. A key observation is the association between the reconfiguration rate of α-synuclein into its natively unfolded state and its aggregation rate. Accordingly, conditions that favor reconfiguration disfavor bimolecular interactions [61], whereas conditions such as high temperature or low pH that slow reconfiguration and reduce diffusibility allow sampling of aggregation-prone disordered conformations (i.e., where hydrophobes are solvent exposed) and thus favor bimolecular interactions and aggregation. Thus, the effect of disrupting interactions between the C-terminus and the NAC region at low pH is to allow the NAC region to instead collapse upon itself (presumably into a more aggregation-prone conformation). These findings provide more evidence that the type and extent of intramolecular α-synuclein interactions are important determinants of intermolecular interactions. These observations also imply that structural probes of compaction, such as FRET and paramagnetic resonance enhancement, and measures of intramolecular diffusion, such as Trp-Cys contact quenching studies, are complementary methods for determining the composition of the unfolded ensemble [61]. Consistent with this hypothesis, it was shown that curcumin, a molecule known to have potent α-synuclein anti-aggregation properties [62, 63], favors α-synuclein reconfiguration to its native state(s) [64].

In summary, the amino acid sequence of α-synuclein dictates local aggregation-determining properties that can rapidly change and interconvert. The exact contribution of each local domain or secondary structure motif to overall aggregation propensity is dependent on multiple factors including, but not limited to, posttranslational modifications, solution conditions, and proximity to other biomolecules. Inhibition of the earliest steps of aggregation at the monomeric level is thus a logical point of therapeutic intervention.

2.4 Toward an Understanding of α-Synuclein Function

Despite intensive investigation, the wild-type function of α-synuclein is not fully understood. α-Synuclein was identified through a presumably unbiased biochemical screen as an inhibitor of phospholipase D. α-Synuclein has been shown to bind to lipid membranes [65, 66], consistent with the structural propensity of the N-terminus to form α-helical structures of the correct size and orientation to interact with phospholipids containing acidic head groups such as sphingomyelin, phosphatidylserine, and phosphatidylglycerol [67]. This interaction may affect the propensity to form aggregates (see Sect. 2.7) and is an example of α-synuclein misfolding causing the loss of wild-type function.

α-Synuclein lacks a transmembrane domain or lipid anchor, so its mechanism of localization to the synaptic bouton is uncertain, though the N-terminal amphipathic repeats likely play a role. α-Synuclein has been demonstrated to have moderate membrane mobility [68], although multiple pools of differing mobility may exist and aggregation may reduce its mobility [69]. Interestingly, the A30P mutation increases mobility, thus supporting the role for N-terminal α-helix in membrane binding that limits mobility [68]. α-Synuclein has a preference for membranes with high curvature, and α-synuclein itself has been shown to bend membranes [70, 71]. Synaptic vesicles are among the smallest biological membranes yet to be classified and therefore exhibit an intrinsic degree of curvature. α-Synuclein binds vesicles [72]. Upon redistribution from the synaptic vesicle to the relatively flat plasma membrane, for example, during exocytosis, α-synuclein disperses, suggesting that curvature of the membrane supports α-synuclein association [73]. α-Synuclein also interacts with membrane microdomains, such as lipid rafts, and this interaction may not be passive, with α-synuclein potentially contributing to raft formation [74].

Its membrane-associated, presynaptic localization and association with the distal reserve pool of synaptic vesicles strongly suggests a role in neurotransmitter release, synaptic function, and plasticity.

Some of the original studies describing α-synuclein function suggested an ability to promote neurotransmitter release [75, 76].

Other studies, however, using α-synuclein knockout mice, demonstrated an inhibitory role for α-synuclein in dopamine vesicle exocytosis (release) and neurotransmission [77, 78], but a supportive role in maintaining the readily releasable vesicle pool, as evidenced by a deficiency in refilling of vesicles and trafficking of vesicles from the reserve pool to the site of release [76, 77]. Further, siRNA knockdown of α-synuclein resulted in decreased availability of the reserve synaptic vesicle pool [75].

Overexpression of α-synuclein, at levels that do not cause neurotoxicity, has supported the assertion of an endogenous inhibitory role by causing decreased catecholamine release in cultured cells [79], including primary hippocampal and midbrain neurons by inducing a reduction in the size of the synaptic vesicle recycling pool [80] and by causing inhibition of intersynaptic vesicle trafficking and exocytosis [81]. Similarly, in PDGFβ-human α-synuclein transgenic mice, a model for DLB [82], the suppression of vesicle exocytosis by α-synuclein is evidenced at the ultrastructural level by enlarged synaptic boutons and convolution of internal membranes connected to the cell surface [83].

A better understanding of the complex role of α-synuclein at the presynaptic terminal resulted from the discovery of a biochemical association and functional interaction of the C-terminus of α-synuclein with vSNARE synaptobrevin 2 [84], the vesicle-bound component of the 4-membered complex that plays a critical role in membrane fusion and vesicle exocytosis leading to neurotransmitter release. As a result of this interaction with vSNARE, α-synuclein promotes assembly of the SNARE complex [84], although overexpression of α-synuclein may inhibit neurotransmitter release through a poorly understood alternative mechanism that involves the α-synuclein-vSNARE (or membrane-bound tSNARE) interaction. Alternatively, α-synuclein may inhibit SNARE-mediated vesicle fusion through direct binding to the adjacent membrane, without an interaction between α-synuclein and any of the SNARE proteins [85]. In any case, the net effect of α-synuclein at or near the SNARE complex is to alter its assembly and thereby modulate neurotransmitter release and vesicle recycling, and the biological effect likely depends on the levels and aggregation state of α-synuclein at the presynaptic terminal.

Genetic evidence links α-synuclein function and the presynaptic chaperone protein cysteine-string protein (CSPα). CSPα knockout mice show progressive synaptic degeneration and death by two months of age that is activity dependent and linked to a deficit in SNARE complex assembly. Interestingly, the CSPα phenotype is worsened by α-synuclein absence and, surprisingly, attenuated by α-synuclein overexpression [86]. By promoting SNARE complex assembly and dampening synaptic activity that contributes to the CSPα degenerative phenotype, α-synuclein may partially compensate from the absence of CSPα [84, 86], suggesting a nonclassical chaperone role for α-synuclein.

Interestingly, knockout of α-synuclein protects against the toxicity of the mitochondrial toxin MPTP/MPP^+ [87], though overexpression may [88] or may not [89] enhance MPTP toxicity, and the role of α-synuclein in facilitating MPTP toxicity is poorly understood. Moreover, α-synuclein expression is upregulated by histone deacetylation, and this confers protection against glutamate-induced excitotoxicity [90]. Taken together, these results suggest that α-synuclein function may be most relevant under conditions of stress.

α-Synuclein also interacts with another synaptic protein, synphilin-1. Although the physiological consequence of this interaction remains uncertain, it appears to promote α-synuclein aggregation, making this interaction a potential therapeutic target (see Sect. 7.4).

α-Synuclein may also interact with the small GTPase3a, a protein known to play a role in synaptic vesicle trafficking and recycling [91, 92]. However, once again, the consequence of this interaction is uncertain given that the normal biological function of α-synuclein itself remains to be fully defined. Nonetheless, recent studies underscore the important role – biological, pathological, or both – that α-synuclein plays in endosomal transport events that depend on the E3 ubiquitin ligase Rsp5/Nedd4, including ER-to-Golgi vesicle trafficking [93].

Whatever the exact physiological function, α-synuclein appears to play an important role in synaptic transmission. Since α-synuclein accumulation, not depletion, is the more disease-relevant state, it is unlikely that therapeutics aimed at limiting overall α-synuclein levels will decrease α-synuclein levels below the critical threshold necessary for wild-type function. Nevertheless, appreciation of the role that α-synuclein plays in synaptic homeostasis should be considered when designing therapeutics that reduce overall α-synuclein levels.

2.5 *Multimeric Intermediates: Defining the Identity and Pathogenicity of Oligomers*

The IUPAC definition of an oligomer is a molecule of intermediate relative molecular mass, the structure of which essentially comprises a small plurality of units derived, actually or conceptually, from molecules of lower relative molecular mass. This definition as applied to the ensemble of α-synuclein oligomers would include not only their general size (i.e., small, medium, large) [94] but also their secondary structure characteristics including β-sheet content and arrangement [95, 96] and α-helical content [40], overall 3-dimensional shape (morphology) [30, 97–100], reactivity with conformation and epitope-specific antibodies [101, 102], and whether they are "on" [103] or "off" [32, 104–106] pathway to fibril formation. In addition, there are "operational definitions" that classify α-synuclein oligomers in terms of their biochemical characteristics or function, including protease resistance [32], effect on cell viability [96, 107, 108], ability to generate reactive oxygen species [32], recruit or activate microglia [109, 110], bind to lipids [40], and form pores in lipid bilayers [103, 111–115]. In some cases, oligomers are formed as a result of an interaction with an exogenous small molecule [116, 117], and some of these oligomers are "off-pathway" to fibril formation, but yet they may share some characteristics with "on-pathway" oligomers (see Sect. 6.2).

Given their transient nature, how has the in vivo toxicity of α-synuclein oligomers been studied and defined? A study using viral expression of α-synuclein in rat brain and human mesencephalic cultured cells demonstrated that oligomers are toxic in vivo [102], likely as a result of aberrant permeabilizing membrane interactions. Interestingly, wild-type α-synuclein demonstrated only modest toxicity, whereas mutant forms, including both naturally occurring (A30P) and engineered mutations (E57K, E35K) that disrupt intramolecular salt bridges and promote oligomerization over fibrillization, were more toxic. These studies demonstrate the superior toxicity of oligomeric (~100 nm, A11 immunoreactive α-synuclein) versus fibrillar α-synuclein, though they also raise some questions about the direct applicability to idiopathic PD where wild-type α-synuclein is the main culprit.

Colla et al. [118] showed that high molecular weight, detergent-insoluble toxic oligomers (defined by reactivity with A11 and FILA-1 antibody) accumulate in the ER and precede full disease pathology in A53T mutant α-synuclein transgenic mice. Interestingly, these toxic oligomers are biochemically heterogeneous, are likely on-pathway to fibril formation, and are also enriched in human PD brains. Salubrinal, an anti-ER stress compound, when administered to the transgenic mice [118], or recombinant adeno-associated virus (rAAV)-A53T α-synuclein-injected rats [119], lessened ER-associated α-synuclein oligomers and disease manifestations [119] including Golgi fragmentation. Dopaminergic neuronal loss in the rats was not prevented by salubrinal treatment, however, consistent with evidence that α-synuclein can activate multiple cell death pathways in addition to ER stress, including mitochondrial dysfunction [120, 121] and changes in gene expression.

Additional evidence pointing to the pathogenicity of oligomeric species comes from a recent report showing that α-synuclein oligomers were toxic when injected into the brains of wild-type mice. These oligomers were morphologically higher-order aggregates of heterogeneous size, such as annular and amorphous protofibrillar oligomers [122], consistent with the heterogeneity described in other studies.

Despite the structural heterogeneity, some features, such as specific β-sheet and β-like structures, may be sufficient, though not necessary for toxicity. For example, rigid β-sheet structure is lacking in engineered mutants with proline mutations in the fibril core. These mutations reduce α-synuclein fibril formation, increase oligomer formation, and increase toxicity in a variety of model systems [123], though the direct relevance of the engineered mutants is uncertain.

α-Synuclein oligomers may also oppose monomeric α-synuclein wild-type function, for example, by opposing the α-synuclein-dependent assembly of SNARE complexes [124]. In addition, aberrant membrane interactions and aggregation at the membrane under certain conditions (see Sect. 2.7) may oppose α-synuclein wild-type function in this critical compartment. Thus, the pathogenic impact of α-synuclein aggregates may be compounded by the interference with its wild-type function.

2.6 α-Synuclein Fibrils: The Pathological Hallmark

Polymerized α-synuclein fibrils, either detected in vivo in Lewy bodies and Lewy neurites, where they are associated with >70 additional cellular components [125], or formed in vitro from purified protein [126], have a remarkably consistent size and morphology, especially when compared to heterogeneous, smaller oligomers. How different are oligomers from fibrils? α-Synuclein fibrils average 10 nm in diameter (range of 8–12 nm), though are much longer, typically greater than 1 μM in length as revealed by X-ray fiber diffraction studies and electron and atomic force microscopy. Fibrils also have a highly ordered cross β-sheet structure, demonstrated by circular dichroism (CD), Fourier transform infrared (FTIR) spectroscopy [31, 126, 127], and NMR, with the core of the fibrils comprised of five β-strands, reminiscent of a five-layered β-sandwich [127], and that run orthogonal to the fiber axis [126]. Critical intermolecular interactions that result in fibril formation likely involve the 100 amino-terminal residues [128]. Fibrils bind characteristically to Thioflavin T/S and Congo red dyes and are microscopically detectable in situ. For example, it is reported that in vivo, α-synuclein filaments are 14–16 nm thick [129] and have lengths ≥1 μM. In contrast, on-pathway (to fibril formation) oligomers are reported to have a Stokes hydrodynamic radius of 7 nm [130, 131] and are much shorter. Spherical oligomers, which may be among the toxic species [107], are even smaller at 2–6 nm in diameter, whereas annular oligomers appear under AFM as a dense, remarkably homogeneous population of globular, mildly flattened oligomers with a narrow size distribution (average diameter of 21 ± 3 nm) and with a less dense interior or a pore-like toroid [106, 112].

Formation of α-synuclein fibrils is an energetically favorable reaction, especially once critical low-frequency nucleation events have occurred. Thus, the process may be kinetically regulated in vivo. For example, an interaction between α-synuclein and a molecule (SERF1a) with a binding constant (K_D) of approximately 10 μM markedly accelerates α-synuclein fibril formation by promoting early nucleation events critical for fibril formation. This is as an example of a biologically relevant, low-affinity interaction [132] that has a profound impact on α-synuclein aggregation and suggests that intervening at the nucleation step is a viable way to limit on-pathway α-synuclein aggregation.

Despite the abundant evidence that various oligomeric species are toxic in vivo, it is highly likely that α-synuclein fibrils are also toxic [133], thereby providing rationale for therapeutic approaches that can prevent their formation and/or lead to their clearance.

2.7 α-Synuclein Membrane Association

The association with biological membranes is a central feature of α-synuclein activity, aggregation propensity, and pathogenicity. As discussed in a recent review [59], α-synuclein binding to membranes as a result of its N-terminal helical propensity can have many consequences, some physiological, such as facilitating membrane-to-membrane interactions or inhibiting aggregation, and some pathological, including promotion of α-synuclein aggregation and pore formation.

Despite its clearly important physiological role at the membrane, many studies [134–140] have revealed a paradoxical effect of membrane interaction by showing that membrane-bound α-synuclein is more aggregation prone relative to the unbound form. Other studies [67], however, showed that helical α-synuclein bound to lipid membranes is less aggregation prone. Whether aggregation occurs may depend on the ratio of lipid or detergent to protein, with higher cytosolic protein concentrations promoting membrane-dependent aggregation and the low protein concentration in artificial vesicles allowing membrane interactions to inhibit aggregation [67, 141]. One mechanism by which lipids or lipid-like molecules may facilitate α-synuclein aggregation is simply by confining the protein to a small and/or two-dimensional surface (the surface of a vesicle or micelle), thereby increasing the effective concentration [142]. Alternatively, certain membranes may induce a partially folded membrane-bound intermediate, similar to what is seen in the presence of intermediate concentrations of fluorinated alcohols such as trifluoroethanol (TFE) [140], where the N-terminal part of the lipid-binding domain is helical and the C-terminal part of the lipid-binding domain is unfolded and unbound [139]. This form may still interact with the membrane, although with low affinity, and it may be that this partially helical intermediate drives aggregation by favoring intermolecular interactions between multiple nearby disordered C-terminal regions of the lipid-binding domain of α-synuclein leading to β-sheet formation [59].

Alternative helical forms of α-synuclein may also promote intermembrane associations that are part of its wild-type function. This may occur as the N-terminal lipid-binding region forms an imperfect "broken" helix upon binding to certain lipids, with amino acids Ala30 to Val37 in a non-helical, more mobile state. Thus, the two separate intramolecular helices within the same N-terminal domain may each bind to a different membrane and allow α-synuclein to serve as a bridge. For example, the membrane topology of fusing or budding vesicles could provide a scaffold for the broken-helix conformation of α-synuclein, thereby bridging these membranes with the plasma membrane [143, 144]. The protein could thus function to anchor the vesicle to the membrane or even possibly to sense the extent of budding/fusion. This model was inspired in part by observations that synuclein expression can lead to a clustering of vesicles in yeast [92] and to a stabilization of docked vesicles in chromaffin cells [79].

Posttranslational modifications (see Sect. 3) also affect α-synuclein membrane interactions. For example, N-terminal acetylation affects membrane binding by increasing helicity of the first 12 residues [145]. Nitration disrupts membrane interactions [146], perhaps because of decreased helicity [146] and/or electrostatic repulsion of the negatively charged nitrotyrosine and the negatively charged phospholipids [147]. This could result in an increase in free α-synuclein concentration and increased propensity to aggregate [146].

3 Posttranslational Modifications of α-Synuclein Affect Its Aggregation

3.1 Oxidation and Nitration

Aging is a primary risk factor for PD. According to the free radical theory of aging, irreversible oxidative changes to critical biomolecules result in decreased cellular function over time, particularly in postmitotic cells with limited capacity for regeneration. Oxidative and nitrative damage to biomolecules accumulates with aging. Therefore, it follows that oxidized α-synuclein may be more prone to forming toxic aggregates. However, while oxidation of α-synuclein does modify the rate, extent, and character of the α-synuclein aggregates, it is not simply that all oxidation results in toxic forms of α-synuclein. In fact, under certain circumstances, oxidation may be protective, by directing α-synuclein to nontoxic, off-pathway stable intermediates. It has even been hypothesized that α-synuclein itself, through its reversible methionine oxidation and reduction driven by methionine sulfoxide reductase (Msr), may serve as a cellular antioxidant [148] by transferring oxygen-based radicals that are generated in modest amounts at cell membranes (i.e., lipid peroxidation products) to more stable and easily contained carbon-centered radicals [149]. This notion is consistent with findings that increased expression of MsrA in *Drosophila* prevents α-synuclein-dependent locomotor and circadian rhythm deficits [150] and MsrA protects against mutant α-synuclein-induced toxicity in cultured cells [151].

A thorough review of the oxidative and nitrative changes that occur in α-synuclein and their effects on aggregation and toxicity is the subject of a recent review [147]. Here, I highlight some topics consistent with the theme that modifications to the α-synuclein monomer have far-reaching effects on its aggregation by affecting the ensemble of intra- and intermolecular interactions that direct its aggregation. As such, these changes may be points of intervention for antioxidants, which are proposed as therapeutic agents for PD, but whose success so far for other neurodegenerative diseases such as Alzheimer's disease is limited [152].

It is proposed that dopaminergic neurons in the substantia nigra are particularly susceptible to neurodegeneration because dopamine is a redox-active molecule. Its enzymatic and nonenzymatic oxidation may yield superoxide, semiquinone radicals, H_2O_2, and other oxidants. Combined with the high iron content in these cells, the environment is favorable for Fenton chemistry that produces the highly reactive hydroxyl radical. Indeed, early reports demonstrated that oxidative conditions in vitro [153] and in cultured cells [154] promote α-synuclein aggregation to amyloid-like structures. Moreover, using primary neurons from transgenic *Drosophila* expressing A30P mutant α-synuclein, it was shown that lessening dopamine-dependent oxidative stress with glutathione treatment or by sequestering dopamine in vesicles reduced toxicity [155]. In addition to iron, metals such as and copper, calcium, aluminum, and zinc have been shown to promote α-synuclein aggregation, perhaps by neutralizing the negative C-terminal charge, thereby promoting a partially folded amyloidogenic intermediate [156, 157], or through other redox-dependent interactions.

Moreover, inflammation, as occurs in PD, can cause oxidative and nitrative stress. Nitrated/oxidized α-synuclein and nigrostriatal neurodegeneration was detected as a result of inflammogen injection into the brains of mice that express human wild-type α-synuclein, but not in α-synuclein knockout mice [158], suggesting that α-synuclein can amplify inflammation into neurodegeneration.

Furthermore, nitrosative stress emanating from the endoplasmic reticulum is shown to be a critical α-synuclein-dependent pathogenic determinant in a cross-platform system comprised of yeast, primary rat neurons, and human neurons differentiated from the induced pluripotent stem cells derived from an A53T mutant α-synuclein PD patient [159].

Nitrated α-synuclein is detected in Lewy bodies [160] and is proposed as a PD biomarker [161]. It is also shown to accumulate with age in the substantia nigra of nonhuman primates [162] and in toxin models of PD [163]. Thus, nitrated α-synuclein may not only serve as a marker of neurodegeneration but may be causally involved in α-synuclein pathogenesis.

Lacking cysteine, α-synuclein is most vulnerable to oxidative and nitrative modification at methionine and tyrosine. Interestingly, oxidation at each of the four α-synuclein methionines to methionine sulfoxide has been shown to retard fibrillization and instead promote unstructured, off-pathway oligomers [148, 164, 165] that are nontoxic to primary cultures of dopaminergic and GABAergic mesencephalic neurons [108].

Similarly, nitration at tyrosine (of which there are four) alters the structural conformation of the monomer in a way that promotes oligomerization and inhibits α-synuclein fibril formation [166], but only in a pure system [167]. On the other hand, when mixed at substoichiometric proportions with unmodified α-synuclein, perhaps more representative of the in vivo situation, nitrated α-synuclein monomers and dimers, with a partially folded conformation, can seed the formation of α-synuclein fibrils [146].

Tyrosine oxidation that alters aggregation also occurs. For example, two α-synuclein-based tyrosyl radicals, formed through reaction with hydroxyl radical, may encounter each other, particularly at high-protein concentration, and form a dityrosine-linked α-synuclein dimer. When a nitrating agent (NO_2–) is included in the reaction, oxidatively cross-linked dimers become nitrated, and this augments the formation of very stable higher molecular weight soluble oligomers. Moreover, preformed α-synuclein fibrils are also greatly stabilized by tyrosine nitration [168]. Nitrative stress increased aggregation of α-synuclein in cultured cells, with a correlation between the number of cells containing elevated peroxynitrite and perinuclear, fibril-like, nitrotyrosine-modified α-synuclein aggregates [169]. Thus, oxidation and nitration are cooperative events that dramatically affect α-synuclein aggregation depending on the degree of stress, the type of modification, and the presence or absence of cellular defenses.

Lipid peroxidation is another common form of oxidative damage with potential pleiotropic effects. Direct modification of α-synuclein by the product of lipid peroxidation, hydroxynonenal (HNE), results in increased production of off-pathway, β-sheet-rich oligomers that are toxic when applied to cultured cells [170], perhaps due in part to their increased capacity to seed further aggregation both within and between cells [171]. These oligomers form following covalent modification of α-synuclein by HNE, likely by Michael addition to Lys and His, although α-synuclein is not cross-linked to itself. The stable intermolecular interactions may result from a combination of hydrophobic, electrostatic, and hydrogen bonding interactions that exemplify the complexity of α-synuclein intermolecular interactions.

Other lipids, for example, docosahexaenoic acid (DHA), affect α-synuclein differently. For example, chemical and biochemical characterization of α-synuclein using ESI-MS in the presence of 5–10-fold excess DHA shows that DHA promotes oxidation of α-synuclein and promotes toxic oligomers containing DHA covalent adducts. At 10:1 molar excess, the oil droplets recruit α-synuclein and promote intermolecular interactions leading to amyloid-like fibrils, though the fibrils appear different than those formed from α-synuclein alone (structurally and chemically). At 50:1, the DHA locks all of the α-synuclein into annular oligomeric forms that are toxic to SH-SY5Y cells. Thus, the type of lipid-promoted oligomers formed depends on the conditions, types of chemical interactions, and the stoichiometry of the interaction [172].

3.2 Dopamine-Modified α-Synuclein: Pathogen or Innocent Metabolic By-Product

α-Synuclein aggregates, including oligomers, that are promoted in vivo under pro-oxidative or nitrative conditions, represent a heterogeneous mixture with different properties, aggregation propensities, and toxicities. This complicates the design of potential therapeutics that aim to limit oxidant-induced aggregates and underscores the importance of case-by-base candidate assessment. Indeed, not all molecules with redox activity will affect α-synuclein in the same way. For example, it is possible that redox-active dopamine, with its electron sink amino group, may be particularly prone to forming reactive quinones, whereas more conjugated compounds may generate less reactive intermediates.

It was initially demonstrated that dopamine modifies α-synuclein through a covalent interaction, likely at Tyr residues, thereby stabilizing off-pathway protofibrils [173] and preventing fibrillization. Recent work, using cyclic voltammetry to demonstrate this dopamine–tyrosine interaction, supports this finding [174]. A similar covalent interaction was proposed for an oxidized quinone form of dopamine and α-synuclein, though the linkage instead occurs at α-synuclein amino groups. The dopamine quinone adducts blocked fibril formation in favor of toxic protofibrillar intermediates [175]. Interestingly, although DA-quinone-induced species were toxic to cultured PC12 cells, toxicity was no greater than that seen with unmodified aged α-synuclein, and sometimes less toxic, depending on incubation time.

Most recently, Nakaso et al. used PC12 cells overexpressing α-synuclein and modulation of TH levels to effect changes in dopamine levels. Decreasing intracellular dopamine blocked vulnerability to α-synuclein, and α-synuclein mutated at oxidizable C-terminal methionines and the YEMPS region was less toxic relative to wild-type α-synuclein, suggesting that dopamine, or its oxidized product, interacts with α-synuclein in the YEMPS region through an oxidative mechanism to promote toxicity. Met127 may be the critical target of oxidative modification, and Tyr125 and Ser129 may enhance that modification. The question of how α-synuclein binds to DA was not directly addressed, although the authors discussed how methionine sulfoxide reductase might reduce oxidized methionines in α-synuclein and lessen its toxicity [176].

Covalent interaction between α-synuclein and dopamine, or a dopamine oxidation product, however, may not be necessary to inhibit α-synuclein aggregation and result in stable oligomers. For example, Zhou et al. assessed the dopamine oxidation product DOPAC and found that at roughly equimolar ratios, its inhibitory effects on α-synuclein fibrillation could be mediated through non-covalent interaction. At high molar excess of DOPAC, concentrations at which DOPAC itself may polymerize, however, the interaction was covalent, as a result of Michael additions [177]. As with dopamine quinone, the DOPAC quinone can lead to α-synuclein Met oxidation, probably as a result of H_2O_2 production as a by-product of DOPAC oxidation. Notably, Met oxidation occurred after the non-covalent interaction between oxidized DOPAC and α-synuclein, suggesting Met oxidation is not a determinant of the initial interaction. Interestingly, of the compounds studied, those compounds with 1,2 dihydroxy groups (vicinal) were not subject to Michael addition at low concentration, whereas compounds containing 1,4 dihydroxy (para OH) were susceptible [177].

How might dopamine-dependent redox reactions affect α-synuclein aggregation? One possibility is that methionine oxidation at the N- and C-termini induced by dopamine inhibits end-to-end association and the resulting oligomers [178]. By using complementary methods, including sedimentation velocity analysis, small-angle X-ray scattering (SAXS), and circular dichroism spectroscopy to study the characteristics of α-synuclein oligomers promoted by DA, these authors demonstrate trimers that consist of overlapping *wormlike monomers*, with no end-to-end associations, and that lack extensive β-sheet structure. They propose that oxidation of the four methionine residues at the C- and N-terminal ends of α-synuclein prevents their end-to-end association and stabilizes oligomers formed by cross-linking with DA-quinone/DA-melanin, which are formed as a result of the redox process, thus inhibiting formation of the β-sheet structure.

Other studies also suggest that non-covalent interactions are responsible for dopamine oxidation products' demonstrated inhibition of α-synuclein aggregation. For example, dopaminochrome was demonstrated to react reversibly with α-synuclein [179], and the ability of dopamine to inhibit wild-type α-synuclein aggregation was demonstrated at substoichiometric concentrations. Furthermore, aged, oxidized dopamine inhibits aggregation of α-synuclein that has been mutated at potential sites of covalent linkage (Met, Tyr, His). These authors propose that dopaminochrome interactions at the YEMPS region induce conformational changes that lessen abundance of classic β-sheet formation in favor of alternative β-pleated sheet-like structures that differ from the β-pleated sheet characteristic of filamentous α-synuclein. Consistent with this finding, Herrera et al. used molecular modeling and in vitro aggregation reactions to show that dopamine interacts non-covalently and nonspecifically through hydrophobic interactions with the YEMPS region to inhibit α-synuclein aggregation [180]. A subsequent study using molecular modeling supports the notion of non-covalent interactions between dopamine and α-synuclein [181], though the in vitro fibrillization experiments in this study did not assess this directly.

In summary, covalent interactions between α-synuclein and dopamine (or its oxidized forms) may be sufficient but not necessary to modify α-synuclein aggregation. In either case, inhibition of fibrillization by dopamine or its metabolites may involve oxidation of α-synuclein Met residues, and dopamine modification of α-synuclein is not necessarily toxic, although this is an ongoing debate.

Importantly, the study of dopamine/α-synuclein interactions has provided valuable information while also eliciting caution in the search for a pharmacophore that could inhibit α-synuclein aggregation. Given the impact of minor structural changes on the activity and mechanism of action, and the range of interactions that dictate α-synuclein aggregation and toxicity, molecules with structures similar to dopamine must be evaluated individually to assess their interaction with α-synuclein and impact on α-synuclein toxicity in vivo.

3.3 Phosphorylation

α-Synuclein contains multiple putative phosphorylation sites, consistent with its extensive phosphorylation in Lewy bodies [58, 60, 182, 183] and the cerebrospinal fluid of PD patients [184]. Phosphorylated α-synuclein is proposed as an early indicator of progressing pathology in specific brain regions [185] and a PD plasma biomarker [186], though plasma levels of phosphorylated α-synuclein vary greatly between individuals.

Despite these observations, the question of whether phosphorylation of α-synuclein plays a causal role in its aggregation and neurotoxicity is a controversial topic that is complicated by a number of factors. First, phosphorylation in vivo occurs primarily at three sites, including Ser129, Ser87, and Tyr125, with some reports indicating that phosphorylation at different sites has opposing effects. Second, reported in vivo outcomes of even a single phosphorylation event, for example, at Ser129, are contradictory. Third, it is not clear if α-synuclein phosphorylation occurs before (early event) or after Lewy body formation (later event), or both. Finally, an extensive network of kinases and phosphatases control α-synuclein phosphorylation.

While a thorough examination of all of these topics is beyond the scope of the review, it is worth addressing some highlights since drug discovery efforts targeting phosphorylation have proven successful for other indications (i.e., cancer) and are a highly active area of PD research.

The consequence of α-synuclein phosphorylation appears to depend on the residue that is phosphorylated and the model system in which it is studied. For example, phosphorylation at Ser87 [60] but not Ser129 [187] interferes with membrane binding.

Some studies conclude that Ser129 phosphorylation of α-synuclein increases aggregation and toxicity in *Drosophila* [188] and toxicity (but not aggregation per se) in cultured cells [189] and rat brains expressing α-synuclein from rAAV vectors [190]. However, the effects of promoting and blocking phosphorylation are not always reciprocal. For example, in the cell-based study [189], blocking α-synuclein phosphorylation with the S129A mutation did not reduce toxicity relative to wild-type α-synuclein, whereas in the rat study [190], the S129A mutant was less toxic but formed larger aggregates. Recently, Sato et al. [191] reported toxicity from viral co-expression of A53T α-synuclein and the GRK6 kinase in rats. These authors report increased Ser129 phosphorylated α-synuclein in vivo without an increase in putatively toxic oligomers, another example of uncoupling of aggregation state and toxicity.

Consistent with the pathogenicity of Ser129 phosphorylation, phosphatase activators that reduced levels of Ser129 phosphorylated α-synuclein also improved motor performance in wild-type α-synuclein transgenic mice [192]. Furthermore, pramipexole, a dopamine agonist with demonstrated neuroprotective properties in model systems, has been shown to decrease phosphorylation at Ser129 by eliciting an inhibitory effect on casein kinase 2, one of the kinases known to phosphorylate α-synuclein [193], although the mechanistic relationship between the multiple activities of pramipexole is not fully understood.

In contrast to the phospho-Ser129-dependent toxicity described in the studies above, others indicate non-phosphorylated forms of α-synuclein may be more deleterious. For example, blocking Ser129 phosphorylation (by using S129A mutant α-synuclein) increases toxicity in vivo in the nigral pathway of rats transduced with rAAV-α-synuclein constructs [194, 195] and causes motor deficits and synaptic deficits in transgenic *C. elegans* (worms). This toxicity may result from increased cytosolic protease-resistant β-sheet-rich aggregates [195] that promote deleterious membrane interactions [196]. In two of these studies [194, 196], mimicking Ser129 phosphorylation (by using the S129D mutant form) was protective, although recent studies [187, 197–199] have cast some doubt on the ability of the S129D and S129E substitutions to reproduce the effect of Ser129 phosphorylation on α-synuclein structure and aggregation in vitro. This disconnect may apply to Ser87 as well, since S87-P but not the phosphomimic S87E induced significant changes in the structure of membrane-bound α-synuclein, suggesting that only authentically phosphorylated α-synuclein may have reduced ability to form α-helical structure when binding to membranes. In the third study, [195] S129D α-synuclein still resulted in some neuronal loss, though it caused less toxicity relative to wild-type (intermediate) and S129A mutant (most toxic) α-synuclein, providing additional evidence that under some conditions, phosphorylation and neurotoxicity of α-synuclein aggregates may not be tightly coupled.

Consistent with non-phosphorylated forms being more deleterious, some reports indicate that α-synuclein phosphorylation may, instead, be protective. For example, Ser87 phosphorylation is reported to inhibit fibrillization by increasing conformation flexibility [60], whereas blocking phosphorylation at this site with the S87A mutation is reported to increase the formation of insoluble protease-resistant, β-sheet-rich aggregates, enhance toxicity, and induce hemi-Parkinsonian motor dysfunction in rAAV-α-synuclein expressing rats [200]. These S87A-dependent deleterious effects are similar to those seen with nigral expression of wild-type α-synuclein and greater than those seen with expression of the phosphomimic S87D mutant α-synuclein [200]. In a similar fashion, Tyr125 phosphorylation may also be protective [201].

At what point in the aggregation process does phosphorylation occur? Detailed structural analysis revealed that Ser87 is inaccessible to casein kinase when α-synuclein is fibrillar, although Ser129 may be phosphorylated after fibril formation has occurred, raising more uncertainty about the cause-and-effect relationship between α-synuclein phosphorylation, aggregation, and neurotoxicity [60]. Moreover, A53T-α-synuclein expressing mice show Ser129 phosphorylated α-synuclein that presents in heterogeneous discrete but overlapping patterns, suggesting that phosphorylation is not a prerequisite for α-synuclein aggregation and may be a marker of advanced pathology [202]. Perhaps phosphorylated α-synuclein is cleared less effectively, consistent with a secondary effect of phosphorylation [189].

In any case, a valid therapeutic approach may involve targeting specific kinases or phosphatases that modulate phosphorylation of specific amino acids. Dual specificity tyrosine-regulated kinase 1a (dyrk1a) is known to phosphorylate α-synuclein at Ser87 [203]. CK-1 and CK-II phosphorylate α-synuclein at Ser129 [204–206]. Polo-like kinases appear to be among the physiological kinases responsible for α-synuclein Ser129 phosphorylation [207, 208] and may play a role in its autophagic clearance [209].

Which kinases are possible therapeutic targets? Perhaps modulating the activity of the G protein-coupled receptor kinase (GRK) 2, another kinase known to phosphorylate α-synuclein at Ser129 [188, 191, 210], or AMP kinase, whose overexpression increases aggregation and causes decreased neurite outgrowth in primary neuronal cultures [211], may be beneficial. Interestingly, α-synuclein phosphorylation [212] and pathogenicity may be mediated, at least in part, by

LRRK2 [213], a kinase that when mutated causes inherited forms of PD and which is the target of extensive drug discovery efforts.

Recently, a cell-based siRNA screen of 711 known or predicted kinases and 206 phosphatases selected for changes in phosphorylated α-syn/total α-syn ratios [214]. This study confirmed that α-synuclein phosphorylation is controlled by an extensive network of kinases and phosphatases. In addition, this study identified a kinase, SMG1, whose knockdown in α-synuclein expressing neuroblastoma cells resulted in increased phosphorylated α-synuclein (and increased total α-synuclein). This counterintuitive negative regulation of α-synuclein phosphorylation by a kinase may be explained, at least in part, by the ability of SMG1 to promote α-synuclein-induced stress granule formation. Reducing the cellular response to α-synuclein-induced stress by knocking down SMG1 attenuates the buffer to further α-synuclein accumulation. Another counterintuitive finding in this study was that knockdown of protein phosphatase 2A, known to dephosphorylate Ser129, instead decreased the phosphorylated α-syn/total α-syn ratio, underscoring the challenge in therapeutic modulation of α-synuclein phosphorylation.

Translation of such findings into in vivo animal models and human proof of concept studies may provide compelling evidence that targeting α-synuclein phosphorylation is a bona fide PD therapeutic approach.

4 Defective Turnover and Clearance as a Consequence of α-Synuclein Aggregation

α-Synuclein has a relatively long half-life that increases with age in neurons [215]. It is normally cleared by both the ubiquitin proteasome system (UPS) [216] and lysosomal pathways such as autophagy [217, 218] (and reviewed in [219]). Even though both are likely important [220], the former may be more relevant under normal physiological conditions, whereas the latter may be more relevant to cellular attempts to clear aggregated pathogenic forms of α-synuclein [218, 221, 222]. Nevertheless, α-synuclein aggregates may resist degradation by both pathways, thereby fueling the vicious cycle that would result in greater aggregate accumulation and generalized defects in protein turnover. For example, α-synuclein aggregates may cause a general impairment of macroautophagy by reducing initiation of autophagy [223] or impairing autophagosome clearance [224]. In addition, mutant or modified (i.e., by dopamine) forms of α-synuclein are shown to impair chaperone-mediated autophagy (CMA) [225, 226]. Moreover, nitrated α-synuclein monomers and oligomers may resist degradation by the lysosome [169]. Additional evidence for a relationship between CMA and α-synuclein toxicity comes from the discovery that the pro-survival transcription factor MEF2D, normally active in the nucleus and degraded by Hsc-70-dependent CMA, instead accumulates in an inactive form in the cytosol when α-synuclein inhibits CMA. Conversely, restoration of MEF2D function in the nucleus attenuates α-synuclein-induced neuronal death [227].

At the proteasome, α-synuclein aggregates are resistant to degradation and impair proteasomal function [228–230]. The net effect of inhibiting these clearance pathways is to increase steady-state levels of both monomeric and aggregated forms, thereby increasing aggregation propensity and neurotoxicity. Thus, it follows that therapeutic approaches aimed at increasing turnover of α-synuclein are actively pursued (see Sect. 7.4).

5 α-Synuclein "Strains" and Prion-Like Transmission

With α-synuclein monomers and oligomers capable of assuming such a heterogeneous collection of 3-dimensional structures of varying function and pathogenicity, it comes as little surprise that even in its most aggregated form, resembling classic amyloid- α-synuclein "strains" exist. While the concept of a "strain" was traditionally used to classify variability of heritable traits, including pathogenicity, at the organismal level, that result from genetic polymorphisms, the term is now commonly used in the context of protein misfolding and aggregation. A number of criteria define a protein aggregate "strain" including how the molecules pack within the aggregate, how well they seed new aggregates, their physical properties, their ability to cross the species barrier, pathophysiological characteristics, latency to disease, and differences in survival rates [231]. For misfolded proteins, this concept was first demonstrated for prions (infectious proteins), a corrupted version of an endogenous cellular protein that is the infectious pathogenic agent in rare but devastating neurological diseases like Creutzfeldt-Jakob disease [232, 233]. The concept has since been applied, in part, to other pathogenic misfolded proteins including Tau, α-synuclein, and Aβ, based partly on evidence that certain strains of one protein aggregate can seed others. For example, α-synuclein aggregates seed Tau aggregation [234], although the prion protein remains the only misfolded protein shown to be transmissible from one organism to another. The cell-to-cell spreading of α-synuclein pathology, in a "prion-like" fashion, was recently covered in two reviews [34, 235] and in a recent publication describing this phenomenon in the context of multiple systems atrophy [236]; hence, in-depth coverage of this topic is beyond the scope of the review. However, a summary and some recent studies are worth noting, particularly in the context of treating PD as a disease that spreads long distances along anatomically connected neural networks, thus raising the possibility that targeting pathogenic extracellular α-synuclein aggregates could be a viable strategy for preventing disease progression.

α-Synuclein pathology spreads in a stereotypical temporal and topological manner, likely starting in the olfactory bulb, the gastrointestinal tract, or both. In central nervous system (CNS), α-synuclein pathology, in general, spreads upward and in a caudo-rostral fashion through specifically affected brain regions, including the nigrostriatal pathway and eventually, in some cases, the cortex. [2, 237].

In vivo evidence for non-cell-autonomous effects of α-synuclein aggregates, i.e., cell-to-cell transmission of aggregates, comes from the demonstration that healthy fetal mesencephalic grafts placed in the diseased striatum of PD patients eventually show evidence of Lewy bodies [238–240].

Numerous in vitro and in vivo studies showed that α-synuclein can be released from one cell, taken up by another, and seed aggregation of endogenous α-synuclein in the recipient cell, with resulting deleterious pathological changes [129, 241–246]. It is likely that this transmission can occur without cell death or membrane leakage of the donor cell, perhaps through exocytosis [247], although cell death and membrane degeneration would accelerate the transmission by releasing a large amount of α-synuclein aggregates. Endocytosis likely plays a role in α-synuclein uptake in the recipient cell [241].

In one study, wild-type human α-synuclein preformed fibrils (pffs), but not monomer, seeded the formation of aggregates in adjacent cultured mouse hippocampal neurons that expressed endogenous levels of wild-type mouse α-synuclein [129]. Seeding was time and concentration dependent and resulted in recruitment of wild-type mouse α-synuclein away from the presynaptic terminal into Lewy neurite-like inclusions, with eventual intracellular propagation of α-synuclein aggregates from axon to cell body to dendrites. The effect of these inclusions was an impairment of coordinated network activity [129].

Most recently it was shown that specific forms of α-synuclein oligomers, including β-sheet-containing dimers [122] and HNE-modified [248] α-synuclein, seed the formation of oligomeric aggregates. For example, when injected into the hippocampi of wild-type mice, the dimer-derived oligomers, but not freshly dissolved α-synuclein or insoluble fibrils, promoted apoptosis [122].

Another study [231] demonstrated two different aggregated α-synuclein forms, depending on solution conditions, that fulfill many of the characteristics of strains, including differences in secondary and quaternary structure (Thioflavin T-reactive fibrils versus Thioflavin T-negative ribbons), the ability to cross-seed unfolded α-synuclein, the ability to bind membranes, and the degree of toxicity. Morphologically the two strains resemble ribbons and fibrils. Under conditions where α-synuclein ribbons form, the soluble precursor of α-synuclein fibrils does not appear to be populated. Thus, even when seeded with preformed fibrils, the result is unfaithful elongation of α-synuclein fibrils into ribbons. However, under solution conditions favorable to fibril formation, the soluble precursor will assume the shape that is consistent with the seeds that are present – i.e., ribbons recruit monomer to form more ribbons and fibrils recruit monomer to form more fibrils. This implies that strain stability is most dependent on the conformational ensembles that the precursor, soluble peptides populate [231] and is consistent with the hypothesis that factors that affect intramolecular compaction, diffusion, and reconfiguration of the monomer will profoundly affect the ultimate rate of aggregation and aggregation state.

6 Experimental Approaches Targeting Misfolded α-Synuclein

Despite the vast amount of knowledge gleaned from the studies described above, many critical questions still remain that are relevant to the development of a PD therapeutic targeting α-synuclein. (1) What is the desired cellular and subcellular concentration of α-synuclein? (2) What are the in vivo toxic species? (3) How does one effectively target a dynamic natively unfolded protein? (4) What are the critical downstream effectors of α-synuclein toxicity? (5) Are α-synuclein-dependent pathologies and functional deficits reversible? Despite having only partial answers to the above questions, a number of approaches are under investigation with the goal of maintaining α-synuclein at physiological levels, limiting toxic aggregates, and lessening downstream effects of misfolded α-synuclein. These approaches include targeting α-synuclein gene expression, targeting posttranslational modifications that affect its toxicity, inhibiting or reversing its aggregation, limiting its cell-to-cell transmission, and enhancing clearance of its toxic forms. The molecules that are being tested in model systems to achieve these aims include nucleic acids, small organic molecules, and biologics.

Despite extensive efforts, α-synuclein-directed therapeutics remain mostly in the discovery phase, with only one approach – the use of antibodies to enhance α-synuclein clearance and limit its transmission, having reached clinical trials (currently Phase 1).

Given the complexity of α-synuclein aggregation, one simple and direct approach that circumvents the nuances of its folding and misfolding is to limit its expression. Support for this approach comes from animal studies in which expression of A53T mutant α-synuclein is conditional and controlled through the use of the tetracycline-regulated "tet-off" promoter and doxycycline administration. In these mice, α-synuclein expression leads to pathology, synaptic structural defects and memory deficits that result from deposition of α-synuclein in the limbic system, including the hippocampus. However, when transgene expression is turned off in adult animals for 3 months (at 9 months of age), 12-month-old mice had cleared α-synuclein pathology in the hippocampus and arrested α-synuclein progression in other regions, and the mice recovered synaptic structural defects. These results support the design of therapeutics aimed at limiting the extent and degree of α-synuclein pathology. The limitation of the applicability of these studies to humans, however, is that intervention would have to occur early in the disease process since extensive α-synuclein pathology and neurodegeneration have already occurred by the time patients are symptomatic. An ideal therapeutic would both inhibit and reverse existing pathology and its downstream effects in patients in whom α-synuclein pathology is both present and progressing. More detailed descriptions of experimental α-synuclein-directed therapeutics are presented next.

6.1 Modulating SNCA (α-Synuclein) Gene Expression

Oligonucleotide-based approaches to modulate gene expression have generated great interest. Originally antisense oligonucleotides were a main focus, but more recently inhibitory RNAs (RNAi) including short hairpin (sh) RNA and related micro(mi)-RNAs have gained favor as means to control gene expression. For example, expression of the SNCA gene is normally modulated, in part, by microRNA-7 [249]. In cultured cells, increasing microRNA-7 decreased α-synuclein and protected against oxidative stress, whereas, conversely, knock-down of microRNA-7 increased α-synuclein levels. In vivo, the mitochondrial complex I inhibitor MPTP decreased microRNA-7, suggesting this could be the mechanism of how α-synuclein is increased in this PD model system. An obvious next step is in vivo enhancement of microRNA-7 as proof of concept that downregulation of α-synuclein elicits improvements in PD-relevant phenotypes.

Another approach to modulate α-synuclein gene expression is to use siRNA directed at the α-synuclein transcript. In this regard, naked (without viral vector) siRNA molecules chemically modified to prevent degradation by exo- and endo-nucleases were directly infused into the left substantia nigra of squirrel monkeys and resulted in reduced levels of α-synuclein mRNA and protein in the infused (left) but not the untreated (right) hemisphere. Quantitative analysis revealed a significant 40–50% suppression of α-synuclein expression in the absence of any adverse effects on dopaminergic neurons [250].

Similarly, α-synuclein-directed ribozymes [251] downregulate α-synuclein gene expression at the RNA level.

SIRT2 is a histone deacetylase (HDAC), and AGK2 is a small molecule that inhibits SIRT2 activity, increases acetylated proteins such as tubulin, and inhibits α-synuclein-induced toxicity in a dose-dependent manner in two in vitro models and transgenic *Drosophila*. Inhibition of SIRT2 promoted the formation of larger α-synuclein aggregates, suggesting the protective effect may result from minimizing aberrant interactions with cellular proteins as a result of decreased surface area [252]. Alternatively, α-synuclein may block protein acetylation, thereby attenuating the expression or activity of proteins that possess endogenous neuroprotective function. If so, blocking SIRT2-dependent deacetylation may compensate for this α-synuclein deleterious effect.

6.2 Small Molecules That Target α-Synuclein Aggregation

Initial discovery of molecules that inhibit α-synuclein aggregation was inspired by other molecules with anti-amyloid activity. It was hypothesized that since different pathogenic amyloidogenic proteins, such as α-synuclein and Aβ of Alzheimer's disease, contain similar structural motifs, such as cross β-sheet, they would respond similarly to many of these molecules. Based in part on this premise, it was demonstrated that small molecules from a number of classes inhibit α-synuclein aggregation at substoichiometric concentrations [117]. Of the 79 compounds tested, belonging to 12 different classes, polyphenols were, as a class, the most potent inhibitors of α-synuclein aggregation as assessed by Thioflavin S fluorescence, electron microscopy, and detergent (sarkosyl) insolubility, consistent with potent inhibition by the polyphenol dopamine (see Sect. 3.2). In addition, compounds belonging to six other classes (phenothiazines, polyene macrolides, porphyrins, rifamycins, Congo red and its derivatives, and terpenoids) also inhibited filament assembly to some extent. Representative structures of compound classes with demonstrated α-synuclein anti-aggregation activity are presented in Fig. 2.

The interaction of many of the inhibitory compounds, including exifone (Fig. 2a), favored the formation of soluble low-to-middle molecular weight oligomeric species over fibrils. To explore binding sites, small molecule inhibition of aggregation was assessed on truncated α-synuclein and compared to full-length α-synuclein. These results showed that exifone and lacmoid (Fig. 2b) partially inhibited aggregation of α-synuclein 1–120, but other compounds were inactive, suggesting that exifone and related molecules may not require, but may still interact with, the C-terminus of α-synuclein. Dopamine, however, absolutely required an intact C-terminus to exert its inhibitory effects, demonstrating that not all polyhydroxylated compounds bind in the same fashion. In contrast to fibrils and oligomeric protofibrils, small-molecule-induced oligomers, purified by size-exclusion chromatography, were nontoxic to neuroblastoma cells, suggesting that these oligomers are off-pathway to fibril formation or that aggregation is halted at a stage preceding formation of a toxic intermediate.

EGCG, a polyphenolic component of green tea (Fig. 2a), is among the best-studied naturally occurring molecules with pharmaceutical-like properties (nutraceuticals) capable of disaggregating and detoxifying preformed α-synuclein aggregates. At equimolar concentrations, EGCG remodels α-synuclein fibrils into unstructured smaller oligomeric species that are seeding incompetent. These oligomers include benign amorphous aggregates that are off-pathway to fibril formation and nontoxic to cultured cells [253]. EGCG also prevents the aggregation of freshly dissolved α-synuclein by binding to the protein backbone at the C-terminus as demonstrated by NMR and quinoprotein staining [254] and diverting the aggregation pathway to nontoxic, detergent stable, unstructured oligomers. EGCG interactions with α-synuclein are complex and likely involve hydrogen bonding, hydrophobic interactions, and possibly covalent interactions and redox cycling. The specific contribution of each interaction is uncertain and likely depends on the species of α-synuclein and the stoichiometry of the reaction.

In addition to green tea components, the main polyphenolic compounds in black tea, theaflavins (Fig. 2a), are potent inhibitors of α-synuclein fibrillogenesis [255]. These polycyclic compounds are formed by oxidation and condensation of the catechin structure. Mixtures of α-synuclein with each of three different theaflavins caused inhibition of formation of toxic α-synuclein oligomers and fibrils in favor of a theaflavin-specific pattern of remodeled nontoxic oligomers, as detected by atomic force microscopy (AFM). The interaction between catechins

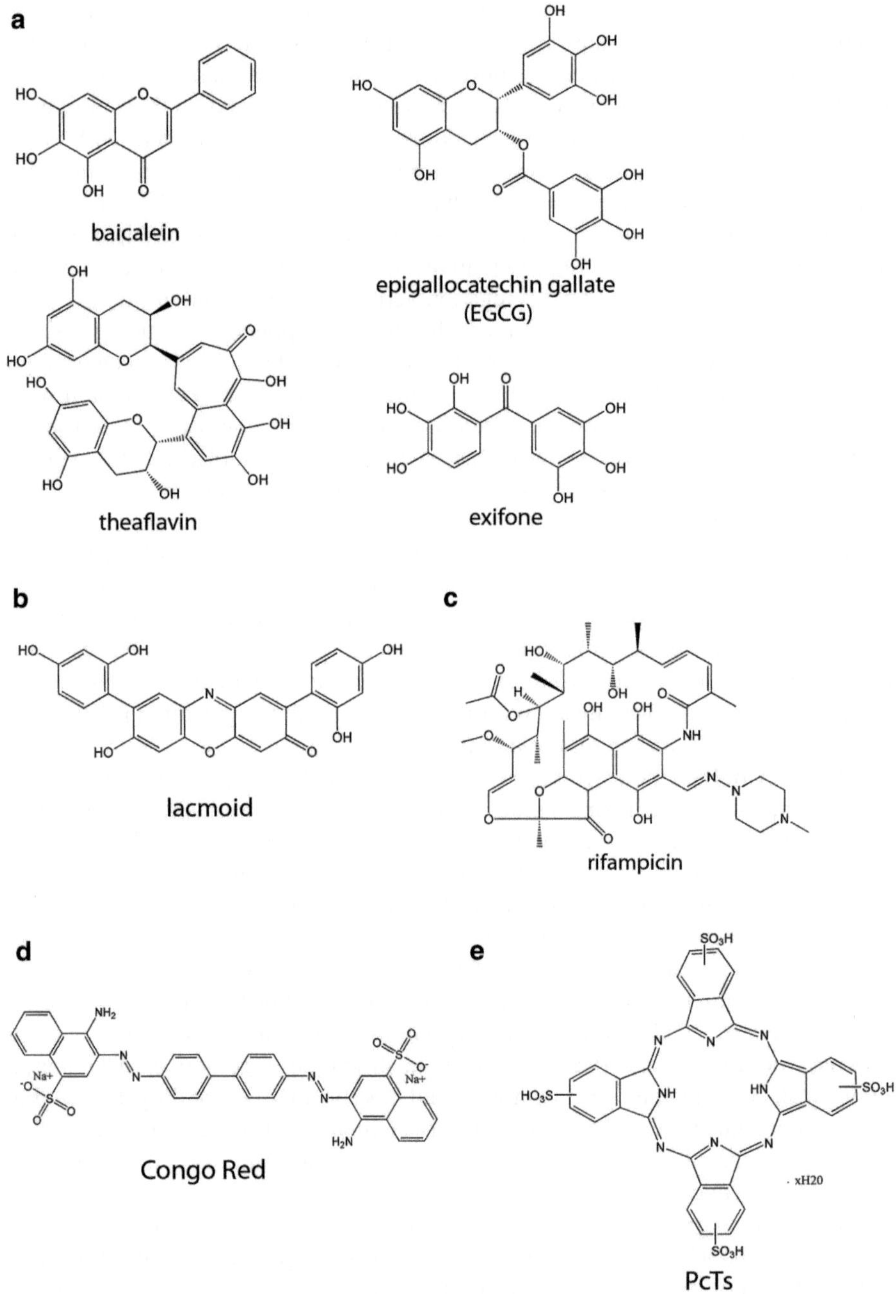

Fig. 2 (continued)

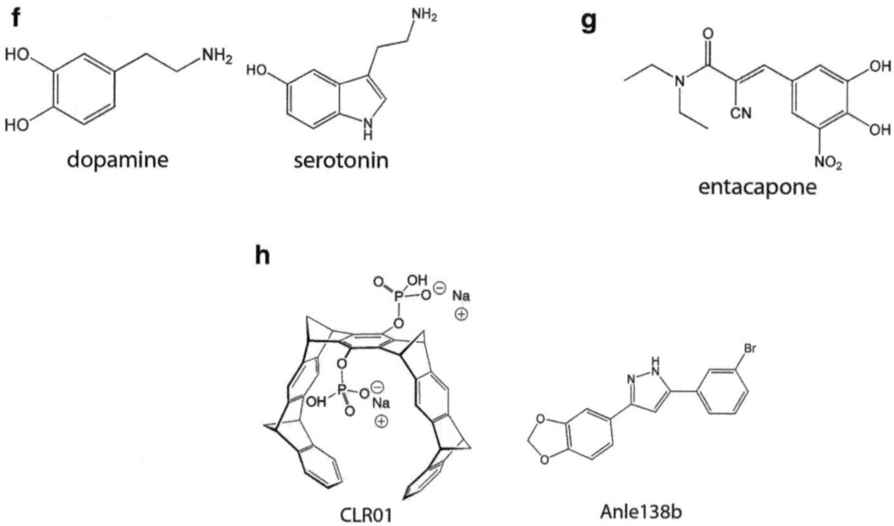

Fig. 2 Representative structures of compound classes with α-synuclein anti-aggregation activity. These structural classes include polyphenols (**a**), phenothiazines (**b**), rifamycin (**c**), Congo red derivatives (**d**), phthalocyanines (**e**), neurotransmitters (**f**), molecules used as symptomatic PD treatments (**g**), and others (**h**) that have demonstrated anti-aggregation activity in a variety of model systems

and amyloidogenic peptides may derive, in part, through presentation of at least one face of multiple phenolic groups in a planar orientation that interrupts amyloid intermolecular hydrogen bonding. These results are consistent with a recent report that polyphenols, including black tea, lessened the effects of exogenous α-synuclein on mitochondrial function [256].

In a related study, 14 polyphenolic compounds and black tea extract were screened using confocal single-molecule fluorescence techniques and short incubation times to assess inhibition and disaggregation of early (small) and intermediate-sized iron-induced on-pathway oligomers [257]. Interestingly, this study defined SAR similar to those in studies assessing fibril formation and disaggregation. For example, multiple compounds, including baicalein (Fig. 2a), EGCG, tannic acid, and NDGA, plus black tea extract, were identified as potent inhibitors and disaggregators of α-synuclein oligomerization, with IC_{50} values in the low μM range. Structurally, aromatic recognition elements are necessary for non-covalent binding to the amyloid core. Moreover, the presence of hydroxyl groups, capable of interfering with intra- and intermolecular hydrogen bonding in the amyloidogenic target, is critical for activity. Although polyphenols commonly have low oral bioavailability, high metabolic turnover, and low blood–brain barrier permeability [258, 259], given their potency, the therapeutic potential is considerable [260].

Another well-studied polyphenol with potent α-synuclein anti-aggregation properties is baicalein, a member of the flavonoid class. Like many polyphenols, flavonoids are abundant in food and potential nutraceuticals. Originally identified as a component of the Chinese medicine, *Scutellaria baicalensis*, baicalein is shown to both potently inhibit α-synuclein fibrillization and disaggregate preformed α-synuclein fibrils, though likely by different mechanisms [116]. Baicalein binds freshly dissolved α-synuclein with a K_D of ~500 nM, most likely at one site, thereby stabilizing a restricted soluble conformation. By electron microscopy and AFM, in time-course experiments, individual oligomers of 15–20 nm in diameter and globular oligomers 32–45 nm in diameter were formed first, followed by larger amorphous protein aggregates. The authors also note the formation of annular oligomers 3.5–5.4 nm in height, 20–23 nm in outer diameter, and 11–14 nm in inner diameter, indicating a diverse oligomeric array. It is proposed that inhibition of aggregation occurs as a result of a covalent interaction between oxidized baicalein quinone and α-synuclein, possibly by Schiff base formation at a lysine side chain, although this adduct is not abundant, requiring purification by RP-HPLC for ESI-MS detection [261]. Interestingly, baicalin, with a covalently attached glucuronide, is active too, suggesting that some forms of metabolized polyphenols, typically highly glucuronidated, methylated, and sulfated in vivo, may retain activity. Purified baicalein-induced oligomers are shown to be nontoxic to membranes, relative to fibrils [262], consistent with other polyphenol-induced α-synuclein oligomers being nontoxic to neuroblastoma cells [117]. Baicalein also decreased fibrillization of E46K-mutant α-synuclein and decreased its toxicity [263].

Baicalein-stabilized oligomers form via a partially folded monomeric intermediate, have β-sheet content between that of monomers and fibrils, and inhibit aggregation of freshly dissolved α-synuclein at substoichiometric proportions [262]. Baicalein also shows modest anti-aggregation activity in a cell-based assay by reducing levels of toxic high molecular weight α-synuclein oligomers and protecting against oligomer-induced toxicity [264]. How might baicalein promote nontoxic oligomers in vitro but show no evidence for such an effect in a cell-based assay? Among the possible explanations for this discrepancy is that the cellular reducing environment is unfavorable for this reaction or, alternatively, the baicalein-induced oligomers are readily cleared in cells.

Although a redox reaction that occurs during the small molecule/protein interaction may be part of its complex mechanism of action, the antioxidant activity of baicalein may not be sufficient for its anti-aggregation activity [261]. For example, following non-covalent binding, perhaps through hydrophobic interactions and pi-pi stacking interactions at specific aromatic amino acids, oxidation to the quinone favors a "reversible" covalent Schiff base formation at Lys 21 (and possibly Lys 23, 34, 43) or a Michael addition at Tyr 39. Consequently, H_2O_2 is produced that catalyzes the reaction between baicalein and α-synuclein, possibly driving Met oxidation. Interestingly, Met oxidation of the protein is not likely a prerequisite for inhibition by baicalein since the H_2O_2 scavenger catalase inhibits Met oxidation, but does not inhibit baicalein quinone formation nor baicalein's inhibitory effects. The authors note that the C-terminal tyrosines may catalyze the inhibitory reaction perhaps as a result of hydrogen bonding with baicalein that consequently alters α-synuclein intramolecular interactions, thereby restricting conformational changes and/or precluding fibril assembly.

The in vitro inhibitory effect of baicalein is consistent with that of several other related flavonoids. For example, 6-HP, myricetin, and eriodictoyl are potent inhibitors of α-synuclein fibrillization, with related but separate mechanisms since the oligomeric species promoted by the flavonoids differ by compound [265]. The most efficient inhibition occurred when select flavonoids were added early in the aggregation process, suggesting that intervention occurs at the nucleation phase and consistent with an elongated lag time detected for α-synuclein fibrillization in the presence of inhibitors. Structure-activity relationship (SAR) analysis confirmed independent findings [257] that vicinal dihydroxyl groups are tightly linked with anti-aggregation potency [265] and more vicinal dihydroxy substituents increase inhibitory potential (i.e., 3 vicinal OH groups are more potent than 2) [265]. Many of the good flavonoid inhibitors of α-synuclein aggregation also disaggregate preformed α-synuclein fibrils [265]. Given the dynamic equilibrium that exists between the different aggregation states, it could be argued that any potent inhibitor of aggregation will result in disaggregation by increasing the "off-rate" for the subunits of the polymerized form.

Another polyphenolic compound, the antibiotic rifampicin (Fig. 2c), is shown to inhibit α-synuclein aggregation and disaggregate preformed fibrils, with low μM potency, and its mechanism of action may be related to quinone formation [266]. Growing evidence suggests that quinone-modified α-synuclein is not necessarily deleterious. For example, a stretch of α-synuclein amino acids (residues 36–46) was covalently modified by pyrroloquinoline quinone (PQQ), EGCG, and baicalein, thereby inhibiting its aggregation. Moreover, PQQ modification of α-synuclein reduced the toxicity of C-terminally truncated α-synuclein (α-syn 119) when the two were coincubated and added to cultured cells. Interestingly, the modified α-synuclein fragment was selective and did not inhibit Aβ aggregation [267]. Another example of nontoxic α-synuclein adducts that promote unstructured, off-pathway oligomers is the consequence of α-synuclein cross-linking with tissue transglutaminase [268].

Zhou et al. [108] further investigated the role of Met oxidation in the formation of stable nontoxic α-synuclein oligomers and propose that the effect of Met oxidation is to strengthen the intramolecular long-range interactions that depend largely on a negatively charged C-terminus. These putative interactions result in a more compact globular structure and prevent formation of the amyloidogenic partially folded intermediate, consequently precluding aggregation, similar to small-molecule-induced oligomers (i.e., baicalein). It is also likely that the addition of more negative charge at the methionine sulfoxide increases electrostatic repulsive interactions. Consistent with the auto-inhibitory role of the C-terminus, the inhibitory effect of Met oxidation is released at low pH, likely by charge neutralization at the C-terminus. Importantly, Met-oxidized α-synuclein oligomers are nontoxic. For example, they do not affect neurotransmitter uptake in primary dopaminergic and GABAergic cells (exposed to 5 and 10 μM Met-oxidized oligomers), whereas positive control HNE-modified α-synuclein oligomers were markedly toxic in this assay [108].

Methionine oxidation, however, does not necessarily imply a covalent small molecule/α-synuclein interaction. For example, the polyphenolic compounds exifone, dopamine (Fig. 2f), and gossypetin inhibit α-synuclein aggregation by promoting stable dimers (and higher-order oligomers) that form through interactions at the α-synuclein N-terminus [269], though dopamine may also interact at the C-terminus [179]. The interactions were determined to be non-covalent, since there was no polyphenol: α-synuclein adducts or α-synuclein cross-linking detected by highly sensitive HPLC/MALDI MS techniques. However, exifone caused the oxidation of α-synuclein at all four methionines. The methionine sulfoxides had only a minor effect on the overall α-synuclein structure as detected by NMR but profoundly impacted aggregation by eliminating fibril formation. Interestingly, Met oxidation was not required for inhibition of fibrillization, since at low inhibitor concentrations, methionine sulfoxide was not detected even though inhibition of fibrillization still occurred. Similar results were found for dopamine and gossypetin.

Given the complex mechanisms of action described for the various molecules in this section, and the structural heterogeneity of the target, it is not surprising that different molecules in the same class may work through different mechanisms to inhibit aggregation of a specific protein. Conversely, a particular molecule may have opposing effects on structurally related targets, for example, by promoting off-pathway oligomers of one target but promoting fibrillization of another [131]. Thus, structure-based predictions regarding mechanism of binding and resulting α-synuclein species are difficult.

6.3 Additional Inhibitory Interactions Between Small Molecules and α-Synuclein

While a single pharmacophore targeting amyloid is unlikely, recently it was proposed that there are similarities in the way small molecules inhibit and disaggregate amyloid fibrils or oligomers with similar structural motifs [270]. Using X-ray crystallography of the steric zipper-containing amyloid structures formed by KLVFFA from Aβ and VQIVYK from Tau, in the presence of small molecule inhibitors including orange G and curcumin, at least two common, class-specific binding mechanisms were identified. In the first, the negatively charged orange G is wedged between the paired β-sheets of the steric zipper, leading to partial opening. The binding is stabilized by packing of the aromatic rings of orange G against the apolar, partially aromatic spine of KLVFFA. Further stabilization arises from salt links between the negatively charged orange G and the positively charged lysines. This stabilization could be a general feature for charged molecules with electrostatic interactions at positively charged Arg and Lys residues. In the second mechanism, the apolar curcumin binds differently, though with a similar result of disrupting the dry interface of the steric zipper by shifting each member of the paired β-sheet. In this structure, curcumin likely lies along the void left by the shifting steric zipper, with the long axis of the curcumin parallel to the fiber axis. The authors suggest that the lack of a site anchoring charge and the nature of the narrow tube binding site running along the β-sheets allows the small molecule freedom to "drift" along the fiber axis. Despite these differences, the common features of orange G and curcumin include binding to the fibers in a similar orientation, with their long axes parallel to the fibril axis and disruption of the β-sheet-paired structure.

Another example of a small molecule with a partially defined amyloid-binding mechanism is phthalocyanine tetrasulfonate (PcTS, Fig. 2e). PcTS when present at molar excess, traps early prefibrillar, amorphous (reduced β-sheet) detergent-sensitive insoluble species that are partially seeding competent. PcTS interacts with the N-terminus of α-synuclein, through aromatic interactions with amino acids including Tyr39 and also via electrostatic interactions. The phalocyanine aromatic system is capable of engaging in pi-pi stacking interactions, thus disrupting intra- and intermolecular interactions, likely involving aromatic residues that favor fibril formation [271]. Indeed, disruption of aromatic interactions may be a common feature of small molecule/amyloid interactions [272].

Despite a focus on the YEMPS region, near the C-terminus, the α-synuclein N-terminus may be a common interaction site for small molecule inhibitors. For example, interactions with chlorazole black E, Congo red (Fig. 2d), lacmoid, PcTS-Cu^{2+}, and rosmarinic acid were examined by NMR spectroscopy to identify α-synuclein sequence elements that are masked by these compounds. Surprisingly, similar α-synuclein interaction sites, encompassing residues 3–18 and 38–51, were detected for all molecules at equimolar small molecule: α-synuclein ratios. At higher ratios, virtually the entire amphiphilic region of α-synuclein (residues 2–92) is affected, revealing the presence of additional, lower affinity interaction sites [273] and consistent with the hypothesis that small organic molecules may drift along the fiber axis.

In addition to dopamine, other neurotransmitters and small molecules that represent symptomatic PD treatments such as selegiline, pergolide, and bromocriptine [274, 275] have α-synuclein anti-aggregation properties. For example, approximately equimolar or greater amounts of serotonin (Fig. 2f) inhibit fibrillization in favor of on-pathway, seeding competent, heterogeneously sized spherical (<50 nm), and granular (100–200 nm) intermediates with partial β-sheet secondary structure. The serotonin/α-synuclein reaction was only partially redox dependent since the reducing agent thiosulfate reduced, but did not eliminate, the potency of serotonin in vitro [276]. Serotonin likely binds directly to intermediates and not monomer, requiring a negatively charged intact α-synuclein C-terminus and a positively charged serotonin amino group.

Moreover, approximately equimolar amounts of norepinephrine, epinephrine, or dopamine or substoichiometric amounts of L-DOPA or baicalein caused sustained inhibition of α-synuclein fibrillization in favor of soluble oligomers as demonstrated by complementary fluorescence-based techniques [277].

Selegiline, a monoamine oxidase B inhibitor and a symptomatic PD treatment, inhibited α-synuclein aggregation at substoichiometric proportions as assessed by fluorescence-based and direct microscopic techniques and resulted in disaggregation of preformed α-synuclein fibrils, consistent with reported neuroprotective effects for this molecule [278, 279].

Entacapone (Fig. 2g) and tolcapone, two nitrocatechol compounds and catechol-*O*-methyltransferase inhibitors, are used with L-DOPA to attenuate its metabolism and prolong its exposure. These two molecules also inhibit the formation of toxic α-synuclein aggregates at substoichiometric proportions in favor of off-pathway nontoxic oligomers [280].

CLR01 (Fig. 2h), described as a molecular tweezer, interferes with α-synuclein aggregation and toxicity in vitro, in cultured cells, and an in vivo α-synuclein zebrafish model [281]. Like other positively charged inhibitors, CLR01 interacts non-covalently with α-synuclein lysine residues. CLR01 interferes with fibril-promoting hydrophobic and electrostatic interactions, likely, in part, by promoting the reconfiguration rate of the monomer [282]. CLR01 likely acts at nucleation and extension and inhibits at stoichiometric proportions, though disaggregation requires molar excess. CLR01 promotes small nontoxic oligomers that migrate as larger oligomers on native gels. CLR01 restores function of the UPS, likely by limiting large aggregates that inhibit its activity. The authors suggest that small molecule inhibitors that bind with moderate affinity are effective because molecular interactions leading to the formation of oligomers are weak; hence, nucleus formation is rare and oligomers are metastable [281].

Curcumin is a primary component in the curry spice turmeric and is widely consumed. Curcumin inhibits α-synuclein aggregation in favor of SDS-resistant oligomers and disaggregates preformed fibrils, both in vitro and in cultured cells with potency in the low μM range [62]. The inhibitory activity of curcumin may be a result of its ability to favor α-synuclein reconfiguration to its native state (s) [64]. Despite low oral bioavailability, due in part to its low aqueous solubility and extensive metabolism, curcumin may pass the blood–brain barrier. Thus, advanced formulations aimed at increasing oral bioavailability such as solid lipid nanoparticles, dendrimers, nanogels (reviewed in [283]), or prodrugs [284] may allow curcumin or a derivative thereof to be used clinically.

Another small molecule with potent anti-aggregation effects was discovered as part of a high-throughput screen, followed by medicinal chemistry optimization of the diphenylpyrazole (DPP) lead to screen for substoichiometric inhibition of prion protein oligomer formation. The lead molecule Anle138b [3-(1,3-benzodioxol-5-yl)-5-(3-bromophenyl)-1*H*-pyrazole] (Fig. 2h) inhibited α-synuclein oligomer accumulation in Thy1-A30P human α-synuclein transgenic mice [285] following oral administration in food starting at 8 weeks of age, using ascending doses, up to 2 × 5 mg per day out to ~69 weeks of age. The authors report good oral bioavailability and brain exposure (up to 125 μM at 10 h post-5 mg dose), modest delays (~10 weeks over 69-week period) in disease onset, modest reductions in α-synuclein aggregates in brain sections, and reductions in oligomers. The molecule does not enhance clearance since overall α-synuclein levels were the same. The authors state that they are advancing this molecule into clinical trials.

Mannitol is an FDA-approved, non-metabolized osmotic diuretic agent. It also has blood–brain-barrier disrupting properties and could possibly be used in combination with other less brain-penetrant drugs to target CNS disorders. Because of its effect on protein stabilization and activity at high concentration that result in molecular crowding effects, it is considered a chemical chaperone. Mannitol demonstrates low potency against α-synuclein aggregation in vitro, with 225 mM as the most effective dose, and a U-shaped dose response curve, with very high concentrations failing to inhibit fibril formation [286]. Despite inhibiting fibril formation, 225 mM did not prevent oligomerization, though very high doses had a modest effect, suggesting different low potency effects on the alternative aggregation pathways. The efficacy of mannitol was tested in vivo using A53T mutant α-synuclein transgenic *Drosophila* (where it reduced brain α-synuclein by 70% and improved behavioral deficits) and in heterozygous transgenic mice overexpressing human wild-type α-synuclein under the murine Thy-1 promoter (mThy1-α-syn/Line 61) [287]. This model has well-characterized α-synuclein pathology in the nigrostriatal pathway, hippocampus and cortex, and behavioral deficits in tests of sensorimotor function. It is widely used to screen potential α-synuclein-directed PD therapeutics. Daily intraperitoneal mannitol injections at 1,000 mg/kg for 4 weeks resulted in beneficial changes in a number of α-synuclein transgene-dependent readouts, including a reduction in detergent-insoluble α-synuclein aggregates in PD-relevant brain regions, reductions in the modest loss of TH-positive projections in the basal ganglia, and decreased α-synuclein-dependent inflammation. The in vivo effects on α-synuclein may result from its demonstrated induction of Hsp-70 rather than direct effects on α-synuclein [286].

Mannosyl glycerate is another putative chemical chaperone. It was shown to inhibit α-synuclein aggregation, in vitro and protect against α-synuclein accumulation, reactive oxygen species production, and toxicity in yeast [288].

Initially discovered for their ability to protect against the toxicity of the amyotrophic lateral sclerosis (ALS)-related protein TDP-43, 8-hydroxyquinolines, specifically HQ-415 and CQ, are shown to have potent rescue activity against α-synuclein foci accumulation and toxicity in yeast and *C. elegans* model systems with potency in the low μM range [289]. The mechanism of action likely involves intracellular metal chelation, though the activity is not likely mediated by reducing metal-dependent reactive oxygen species. Future studies should elucidate if there are in vivo effects in higher-order species.

Chlorogenic acid, a major polyphenol found in coffee, was originally identified as a weak in vitro α-synuclein aggregation inhibitor [117]. Subsequent work showed that chlorogenic acid prevents oxidized dopamine from promoting toxic α-synuclein oligomers and also directly protects against α-synuclein toxicity in cultured cells at low μM concentrations [290].

6.4 *Biologics: Peptides and the Promise of Immunotherapy*

Therapeutic peptides have gained attention in recent years as a result of successes in treating a variety of conditions, particularly those involving hormonal deficiencies or imbalances. Despite potential shortcomings, such as instability and difficult in delivery, peptides may offer advantages for treating CNS disorders such as PD including target specificity, low toxicity, and limited drug–drug interactions.

Initially, peptides targeting α-synuclein incorporated the hydrophobic central region of α-synuclein and contained *N*-methylated amino acids. These peptides were designed such that when incorporated into growing aggregates, intermolecular hydrogen bonding, and thus subsequent self-association, would be interrupted [291]. Another peptide-based approach to inhibiting α-synuclein fibrils in favor of unstructured, nontoxic states is the use of inhibitory hairpin peptides that bear no sequence similarity to α-synuclein [292]. These ~16 amino acid long hairpin peptides contain exposed hydrophobic sites such as Tyr and Trp substitutions that facilitate intramolecular folding. The mechanism of action in the presence of α-synuclein is that the prestructured hairpin-containing strands facilitate intermolecular association in α-synuclein that precludes formation of mature oligomers and fibrils in favor of insoluble amorphous precipitates that stain less extensively with Congo red and do not exhibit red-green birefringence when viewed under polarized light.

The synuclein family member β-synuclein is shown to have potent α-synuclein anti-aggregation activity in vivo [99, 293] (reviewed in [294]). These results are consistent with substoichiometric inhibition of α-synuclein aggregation in vitro [295], likely through the formation of stable non-propagating heterodimers [99]. These results lead to the generation of a β-synuclein-derived peptide library, though peptides derived from this library did not result in beneficial effects on α-synuclein in vivo (reviewed in [294]).

Additional peptide-based approaches have been developed to interfere with specific aspects of the fibrillization process. For example, peptides and peptidomimetics such as KKDQLGK and peptidomimetics that bind to the C-terminus of α-synuclein and prevent head-to-head propagation at membranes are being tested [296]. While overall drugability characteristics may require optimization for peripheral administration, proof of concept in targeting brain α-synuclein has been established by intraventricular injections.

Immunotherapy for neurodegenerative disease was once thought to be applicable primarily to Alzheimer's disease, in which the key pathogenic determinant, Aβ, is excreted into the extracellular space and thus a more accessible target for antibody-mediated clearance. Aβ-directed immunotherapies are currently in Phase 3 clinical trials. Discoveries in the last 5 years, however, have established α-synuclein as both an intracellular and extracellular target that is secreted via exocytosis and can penetrate recipient cells and accumulate at the plasma membrane. These findings provide rationale for the development of immunotherapies for synucleinopathies (reviewed in [297]).

α-Synuclein-directed immunotherapy approaches were first tested preclinically in PDGFβ-human α-synuclein transgenic mice [298]. Following immunization with recombinant human α-synuclein, these mice displayed reductions in α-synuclein in the neuronal cell body and synapse, including reductions in membrane-associated oligomeric α-synuclein species and attenuated neurodegeneration. The antibodies generated in response to immunization were shown to interact with α-synuclein in the lysosome, suggesting that immunization stimulated α-synuclein clearance by lysosomal activation. Importantly, immunization did not stimulate a neuroinflammatory response in these mice.

Given the concerns about potential side effects associated with active vaccination, such as vasculitis and autoimmune responses that have been noted in humans during active immunization clinical trials, emphasis is currently being placed on passive immunization paradigms that utilize administration of antibodies instead of antigens.

In active immunization paradigms, it was noted that many of the high affinity α-synuclein antibodies were directed against the C-terminus, perhaps because this portion of the protein is exposed to the extracellular space upon α-synuclein oligomer binding to and partial penetration of the plasma membrane [299]. Passive immunization with antibody 9E4, directed against amino acids 118–126 of α-synuclein, results in clearance of α-synuclein in the neocortex and hippocampus of PDGFβ-human α-synuclein transgenic mice. These mice also displayed improvement in α-synuclein transgene-dependent deficits in the Morris water maze, improved synaptic pathology, and reductions in calpain-cleaved α-synuclein. Importantly, 9E4 immunization did not perturb the microvasculature [300].

How do peripherally administered α-synuclein-directed antibodies lead to its clearance? Circulating antibodies may recognize aggregated α-synuclein at the plasma membrane. Antibody binding may then, in turn, stimulate receptor-mediated endocytosis and degradation [299, 300] likely by enhancing autophagy in the lysosome [300].

In addition to cell surface recognition, antibodies may recognize oligomeric α-synuclein in the extracellular space and block transmission to neighboring cells [248], including trans-synaptic spreading, and extracellular propagation from cell body to cell body, as well as transmission to grafted neural stem cells and endogenous astrocytes. α-Synuclein uptake by the latter is associated with the release of cytokines, including IL-1α, IL-1β, and IL-6 [243], suggesting that α-synuclein antibodies can block an α-synuclein-dependent proinflammatory cascade. Moreover, antibodies may stimulate physiological microglial activation leading to increased clearance and, in turn, attenuate the proinflammatory response initiated by these cells.

Preventive immunization approaches may be more efficacious and associated with fewer side effects than therapeutic immunization. Nonetheless, the first PD Phase 1 clinical trial utilizing active immunization against α-synuclein was initiated in 2012 by Austria-based AFFiRiS AG. The vaccine, AFFITOPE PD01, consists of a peptide-carrier conjugate and is formulated with aluminum hydroxide as the immunological adjuvant [301]. Though the sequence of the peptide is not published, it is reported that the peptides used in the vaccine are small enough to avoid an α-synuclein-specific T-cell-mediated autoimmune response and also designed to exclude cross-reactivity with β-synuclein. Immunization with AFFITOPE PD01 reduced the level of cerebral α-synuclein in synucleinopathy mouse models and ameliorated α-synuclein-triggered neuropathological alterations such as neuronal cell loss and dendritic density loss [297]. This represents the first active immunization study to be tested clinically in PD.

The use of single-chain variable domain antibody fragments (scFv) is another antibody-based approach to inhibit α-synuclein aggregation, toxicity, and extracellular propagation. One such antibody, D5 scFv was shown to be a potent inhibitor of α-synuclein aggregation in vitro, likely by binding to an early oligomeric form. This antibody also protects cells from α-synuclein-dependent toxicity, possibly by interfering with deleterious oligomeric α-synuclein/cell membrane interactions [302]. Another antibody, syn-10H scFv, was shown to bind to a later stage oligomer, inhibit further aggregation in vitro, and neutralize α-synuclein-induced toxicity in cultured cells [30]. It is proposed that the therapeutic potential of these antibodies can be asserted either extracellularly or via intracellular expression (known as intrabodies). Intrabodies have been used to neutralize toxic effects of different pathogenic agents, including α-synuclein [303].

7 Molecules That Inhibit Downstream Toxicity of α-Synuclein

7.1 ER-to-Golgi and Other Vesicular Trafficking Pathways

N-aryl benzimidazole (Fig. 3a) was shown to suppress α-synuclein toxicity in a multiplatform system including yeast, *C. elegans*, and rat primary neurons expressing α-synuclein from a viral vector with potency in the low μM range. These effects likely result from rescue of trafficking defects that are dependent on the E3 ubiquitin ligase Rsp5/Nedd4 [93] and are related to nitrosative stress [159]. Consistent with an earlier report [304], these studies further define the role of α-synuclein as a "hub" protein, with effects on multiple cellular trafficking-related pathways. These include blocking early ER-to-Golgi events, as well as later exocytic events, early endocytosis, endosomal transport, retrograde transport, autophagy, lysosome vacuole function, ER stress, and ER-associated protein degradation.

Most recently, the benzene sulfonamide-containing compound ELN484228 was shown to block α-synuclein-dependent vesicular dysfunction and the resulting impairments in phagocytosis in cellular models, possibly by blocking redistribution of α-synuclein to the phagocytic cup [305]. ELN484228 also improves survival of primary midbrain neurons overexpressing A53T mutant α-synuclein, with potency in the μM range. ELN484228 does not alter α-synuclein aggregation, although it is predicted in silico to interact with α-synuclein monomer in a putative binding pocket.

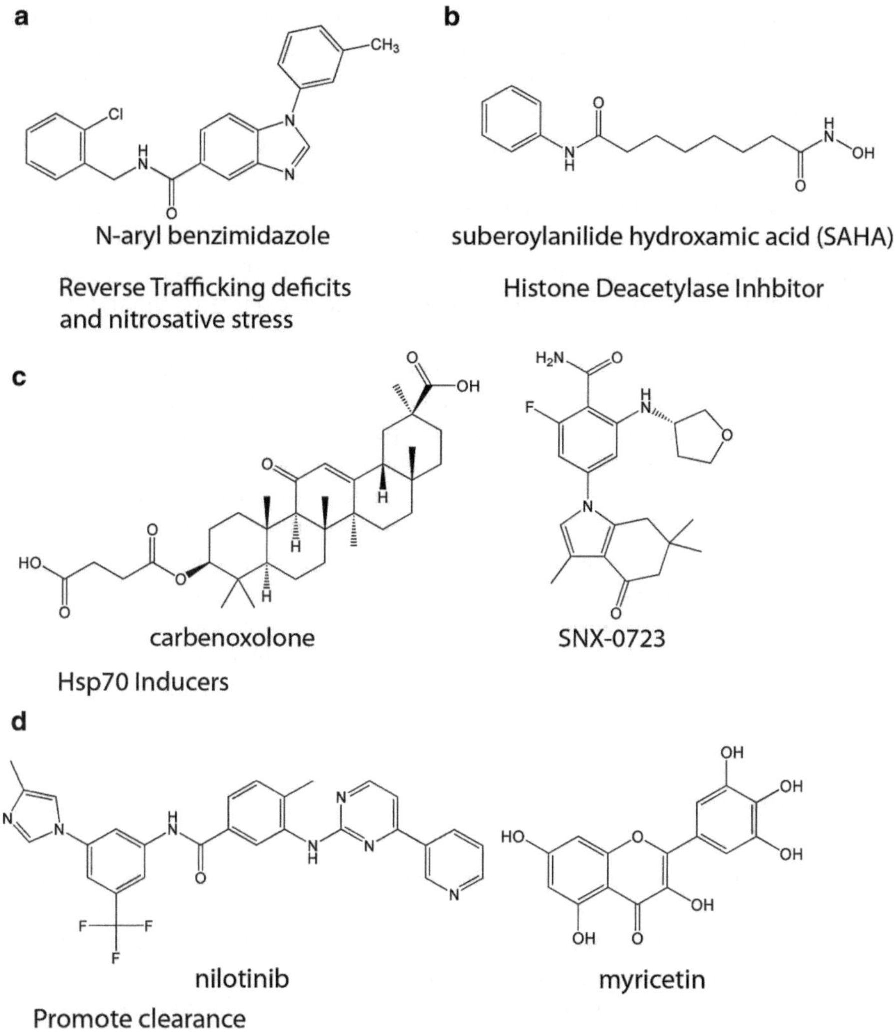

Fig. 3 Structures of select compounds that inhibit downstream toxicity of α-synuclein aggregates. These molecules down-modulate cellular dysfunction and/or stress in the ER/Golgi compartment (**a**), inhibit histone deacetylase activity (**b**), upregulate molecular chaperones such as Hsp70 (**c**), or promote clearance of toxic α-synuclein aggregates via the autophagy or proteasome pathways (**d**)

7.2 Blocking Deleterious Effects in the Nucleus

Using cultured cells and transgenic *Drosophila*, it was shown that α-synuclein that is targeted to the nucleus is more neurotoxic than α-synuclein that is excluded from the nucleus. α-Synuclein may act to modulate gene expression by interacting directly with histones and blocking their acetylation. For example, dopaminergic loss induced by nuclear α-synuclein was reduced by the histone deacetylase inhibitors sodium butyrate and suberoylanilide hydroxamic acid (SAHA, Fig. 3b) [306]. HDAC inhibitors did not affect overall levels of α-synuclein. Valproic acid, another histone deacetylation inhibitor, was shown to protect against rotenone-induced monoubiquitinated α-synuclein nuclear localization and toxicity [307]. In contrast, another study [90] demonstrated that α-synuclein is induced by valproic acid in rat neurons and that this induction was protective against glutamate-induced neurotoxicity. Therefore, modulation of the deleterious effects of α-synuclein on histone acetylation-dependent gene expression alterations may prove challenging.

Other epigenetic factors such as DNA methylation may also mediate α-synuclein expression and toxicity and therefore be a potential therapeutic target. For example, human postmortem brains from PD and DLB patients showed reduction in levels of nuclear DNA methyltransferase 1 (Dnmt1) and consequential reduced DNA methylation [308]. Physical association of α-synuclein with Dnmt1 might mediate retention of Dnmt1 in the cytoplasm. Hypomethylation of DNA could in turn increase expression of α-synuclein, since the SNCA gene is shown to be regulated by methylation at intron 1, with decreased methylation detected in PD patients [309, 310]. Hypomethylation could also alter gene expression of factors that promote neurodegeneration directly or by counteracting other protective molecules. Such factors, if identified, could be therapeutic targets.

7.3 Mitochondria

Mitochondrial dysfunction, in particular electron transport chain complex I deficiency, is observed in a majority of PD patients [311]. Numerous recessively inherited forms of PD involve loss of function in proteins directly involved in mitochondrial maintenance, dynamics, or homeostasis, including Parkin, Pink1, and DJ-1 [312, 313]. α-Synuclein may have a normal role in the mitochondria, such as maintenance of mitochondrial calcium homeostasis [314], whereas pathogenic α-synuclein may target mitochondria [121, 315–318] through direct interactions in a conformation or aggregation state-dependent fashion [319–321] or indirectly through activation of cell death pathways in which mitochondria play a critical role [120, 228].

Mitochondrial oxidative stress is a potential link between α-synuclein aggregation, impaired neuronal function, and cell death [320, 322–325]. Recent studies showed that α-synuclein, like the mitochondrial toxin rotenone, directly inhibits complex I [316, 326] causing increased oxidative stress. In addition, α-synuclein may mediate the effects of mitochondrial complex I inhibitors such as MPTP and rotenone [87]. Interestingly, α-synuclein-dependent reactive oxygen species emanating from the mitochondria and activation of the mitochondrial permeability transition pore may play a critical role in generating a mitochondria to nucleus pro-death signal via the release of the proapoptotic mitochondrial endonuclease G [325].

A53T mutant α-synuclein increases cell death by increasing autophagy of mitochondria [327], whereas another report demonstrated that wild-type α-synuclein promotes mitochondrial fission that results in a decline in mitochondrial function and cell death [328]. The A53T mutation may also impair a wild-type α-synuclein function at the mitochondria-associated ER membrane, thereby causing reduced apposition of mitochondria to ER and increased mitochondrial fragmentation [329].

Therefore, molecules that target α-synuclein/mitochondrial interactions, or lessen the effect of α-synuclein on the mitochondria, may be of therapeutic benefit.

7.4 Enhancing Clearance of Toxic α-Synuclein Aggregates

As described in Sect. 4, α-synuclein is cleared by both the UPS and autophagy. Thus, enhancing clearance of pathogenic α-synuclein species via modulation of these clearance mechanisms with small molecules and biologics is an active area of research.

For example, in addition to their potential direct effects on α-synuclein aggregation, the polyphenols myricetin (Fig. 3d) and purpurogallin blocked the α-synuclein-dependent proteasome inhibition [230], thereby potentially increasing clearance of toxic α-synuclein aggregates and raising the question of whether other common polyphenols have a similar mechanism of action, in addition to their direct anti-aggregation activities.

Molecular chaperones have evolved to provide quality control in the face of protein misfolding, either by recognizing misfolded proteins and preventing their aggregation or, in the case of protein-remodeling factors, by directly resolving protein aggregates that have already formed. Heat shock proteins (HSP) have been shown to fulfill both of these roles and thus may constitute promising therapeutic strategies.

The best understood of the molecular chaperones that modulate α-synuclein aggregation and toxicity is Hsp70, a protein that is upregulated as part of a protective stress response to α-synuclein misfolding and aggregation. Hsp70 inhibits α-synuclein aggregation in vitro at substoichiometric proportions through transient interactions with a pre-fibrillar form at the central NAC region that promotes monomer [330] and/or redirection to amorphous nontoxic aggregates [331]. In vivo, Hsp70 may direct misfolded species to degradation by the proteasome or the lysosome, although specific putatively toxic oligomeric forms of α-synuclein are shown to inhibit the Hsp70 chaperone system. Taken together, these results suggest that α-synuclein toxicity may result from reduced output from the chaperone machinery, thereby affecting α-synuclein and other improperly folded proteins and leading to cellular toxicity [106].

Hsp70 overexpression reduces α-synuclein pathology in transgenic mice [332] and reduces α-synuclein toxicity in transgenic *Drosophila* [333] and in cultured cells [252], including secreted α-synuclein oligomers that may be transmitted from cell to cell [334]. Most recently, chemical induction of Hsp70 levels with carbenoxolone (CBX, Fig. 3c), a glycyrrhizic acid derivative, was sufficient to reduce α-synuclein aggregation and related toxicity in cultured cells, thus favoring pharmacological manipulation of this pathway [335].

Despite having no mammalian ortholog, yeast Hsp 104 was shown to ameliorate α-synuclein expression-dependent neurodegenerative changes in the nigrostriatal pathway of A30P mutant α-synuclein expressing rats, possibly as a result of an Hsp104-dependent reduction in the percentage of cells containing phosphorylated α-synuclein aggregates [336]. Moreover, similar to small molecules such as EGCG described above, and unlike Hsp70, Hsp104 has both inhibitory and aggregate remodeling activity. For example, Hsp104 was able to inhibit α-synuclein aggregation in vitro, even at substoichiometric proportions, remodel preformed oligomers of A30P mutant α-synuclein, and remodel fibrils derived from wild-type, A30P, or A53T mutant α-synuclein. Furthermore, Hsp104 was shown to cooperate with Hsp70 in vitro [336].

Hsp90 is another chaperone studied in the context of α-synuclein aggregation, although it may promote rather than retard aggregation. For example, in one study, Hsp90 was shown to bind to mature monomeric α-synuclein in an ATP-dependent manner, abolish its interaction with small unilamellar vesicles, and promote fibril formation over oligomer formation [337]. In the absence of ATP, however, on-pathway oligomer formation was promoted. The toxicity of these oligomeric species was not assessed, though the results suggest that downregulation of Hsp90 activity could be beneficial. Consistent with this hypothesis, brain-permeable Hsp90 inhibitors such as SNX-0723 (Fig. 3c) were shown to upregulate Hsp70 expression, inhibit formation of high molecular weight toxic oligomers, and rescue α-synuclein toxicity in cultured cells [338].

In contrast, however, a potentially beneficial role for Hsp90 was recently put forth. In this study, Hsp90 was shown to inhibit the aggregation of the A53T mutant α-synuclein. Using a variety of biophysical techniques in vitro, these results demonstrated that Hsp90 redirects α-synuclein aggregation in an ATP-independent manner by forming a strong complex with transiently populated toxic oligomeric α-synuclein species and rendering the oligomers nontoxic to cultured cells [339].

Taken together, these results demonstrate that Hsp90 interactions with α-synuclein and the resulting effect on α-synuclein toxicity may depend on concomitant changes in Hsp70 levels and on ATP availability. Thus, additional studies will be necessary to determine the potential of inhibiting Hsp90 as a PD therapeutic.

α-Synuclein is believed to activate the protein tyrosine kinase cAbl through phosphorylation, and this activation is hypothesized to promote α-synuclein-dependent neurodegenerative changes, consistent with increased phosphorylated Abl levels in the striatum of PD patients and in mouse synucleinopathy models [340]. Conversely, injection of lentiviral vectors driving expression of Abl into the substantia nigra of wild-type mice resulted in a marked increase in α-synuclein monomer and high molecular weight oligomers, supporting a positive association between Abl and α-synuclein levels in brain.

Nilotinib (Fig. 3d) is a brain-penetrant Abl inhibitor approved by the FDA for the treatment of leukemia. Daily intraperitoneal injections of nilotinib for 3 weeks at 10 mg/kg decreased total brain α-synuclein levels in prion promoter (Prp)-A53T human α-synuclein transgenic mice. Decreased α-synuclein was accompanied by an increase in the autophagy markers beclin-1 and Atg12 and a decrease in the autophagosome marker LC3, suggesting that nilotinib inhibition of Abl promotes autophagic clearance of α-synuclein. Injected nilotinib also markedly decreased α-synuclein accumulation in the substantia nigra of an α-synuclein lentiviral vector mouse model, attenuated the α-synuclein-dependent increase in proapoptotic caspase-3, and partially rescued the α-synuclein-dependent loss of tyrosine hydroxylase-positive neurons in the SN [340]. A follow-up study demonstrated that nilotinib and the related compound bosutinib promote parkin activity and its interaction with Beclin-1, thereby promoting proteasomal and autophagic clearance of α-synuclein [341]. These results, demonstrating that pharmacological activation of autophagy can mediate clearance of α-synuclein, are consistent with previous results demonstrating that genetically increasing autophagy via overexpression of beclin-1 [342] or Atg7 [223] in cultured cells and the brains of PDGFβ-human α-synuclein transgenic mice enhances clearance of brain α-synuclein. These results are also consistent with studies showing that rapamycin or rapamycin analogs are capable of reducing the accumulation of α-synuclein oligomers via the autophagy pathway [343]. Rapamycin inhibits phosphorylation of mTor (target of rapamycin) and activates autophagy. However, rapamycin, which is poorly brain penetrant, also downregulates protein synthesis and cell proliferation, potentially complicating its therapeutic use.

α-Synuclein aggregation may also have indirect effects on autophagy by sequestration of autophagy-related transcription factors. For example, overexpression and aggregation of α-synuclein in nigral neurons result in the cytoplasmic sequestration of the autophagy-regulating transcription factor EB (TFEB) and, as a consequence, impaired autophagy. Conversely, TFEB gene transfer, using intranigral injections of rAAV vectors driving TFEB expression, induced autophagy, similar to transfer of beclin-1. TFEB expression caused reductions in high molecular weight α-synuclein oligomers and protected against α-synuclein-dependent dopaminergic deficits and neurodegenerative changes [344].

UPS involvement in PD was supported by the discovery of mutations in parkin, a ubiquitin-protein ligase, and UCH-L1, a deubiquitinating enzyme involved in ubiquitin recycling. Mutations in parkin that result in partial or complete loss of parkin activity cause autosomal recessive juvenile onset PD. Parkin was shown to interact with glycosylated α-synuclein in Lewy bodies [345]. Parkin likely plays a role in α-synuclein clearance and Lewy body formation, consistent with a study demonstrating that lentiviral-mediated expression of parkin in the substantia nigra protects dopamine neurons against A30P α-synuclein-induced neurotoxicity and increases the number of phosphorylated α-synuclein inclusions [16]. In addition to illustrating the role that parkin may play in α-synuclein-dependent toxicity, this study also exemplifies the potential benefit of sequestering α-synuclein in highly insoluble inclusions. Thus, parkin may detoxify proteins in both the soluble and aggregated forms, although parkin itself may not offer an obvious target for therapeutic intervention.

Parkin is also shown to interact with and ubiquitinate synphilin-1 [346]. Synphilin-1 binds to α-synuclein at its N-terminus [347] and prevents its degradation in the proteasome [348]. Aggresomes are formed by convergence of smaller aggregates via microtubule-based transport, cleared by autophagy and thus neuroprotective relative to smaller, diffusible aggregates. Consistent with this hypothesis, enhancing formation of α-synuclein containing aggresomes by modulating synphilin-1 [349] enhanced autophagic clearance [350] and was cytoprotective in cultured cells [351], *Drosophila* [352], and transgenic mice [353]. Other studies in yeast [354] and mice [355], however, have shown that synphilin-1 does not neutralize α-synuclein toxicity in vivo and may instead enhance it. Thus, while an intriguing target, the therapeutic potential of modulating synphilin-1 remains to be determined.

Neurosin is a serine protease that degrades monomeric and oligomeric α-synuclein, resulting in nontoxic fragments. Neurosin levels are reduced in the brains of both DLB patients and PDGFβ-human α-synuclein transgenic mice. Exogenous delivery of neurosin, using intracerebral injection of lentiviral vectors driving its expression, can reduce the brain accumulation of toxic α-synuclein in the mouse model. For example, neurosin delivery attenuated the α-synuclein-dependent reductions in microtubule-associated protein (MAP)-2 and synaptophysin at the synaptic terminal and decreased α-synuclein-induced microgliosis [356].

Mutations in the glucocerebrosidase (GCase) gene that cause the lysosomal storage disorder known as Gaucher's disease result in the highest genetic risk for developing PD and DLB, even among non-symptomatic heterozygous carriers [357, 358]. Indeed, a decrease in glucosylceramidase activity or the presence of a mutant gene can increase amyloid formation of purified α-synuclein by stabilizing soluble on-pathway oligomeric intermediates [359]. Moreover, transgenic mice with GCase mutations demonstrate α-synuclein pathology in the CNS [360]. In addition, PD patients without GBA1 gene mutations have decreased GCase activity [361] consistent with the finding that α-synuclein can inhibit glucocerebrosidase activity in the lysosome [359].

rAAV-mediated expression of glucocerebrosidase in the CNS of symptomatic homozygous mutant GCase mice rescued the aberrant accumulation of the toxic lipid glucosylsphingosine and reduced the levels of ubiquitin, tau, and Proteinase K-resistant α-synuclein aggregates. In addition, hippocampal expression of glucocerebrosidase in these mice also reversed their cognitive impairment even when expression is started at a postsymptomatic age. In regards to effects of GCase expression on α-synuclein, overexpression of GCase in A53T mutant α-synuclein transgenic mice reduced the levels of soluble α-synuclein, suggesting that increasing GCase activity can serve as a therapeutic strategy for both GBA1-related and unrelated α-synucleinopathies [360, 362]. Regardless of GBA1 genetic background, it is postulated that increasing lysosomal GCase activity prevents the accumulation of toxic metabolites, thus improving lysosomal function leading to improved clearance of other toxic proteins including α-synuclein aggregates. Decreasing toxic α-synuclein aggregates may, in turn, allow proper trafficking of GCase to the lysosome and support its enzymatic activity [359, 362, 363]. Taken together these results suggest that augmenting glucocerebrosidase activity is a viable PD therapeutic approach.

Cathepsin D is another lysosomal enzyme that is now considered one of the main lysosomal enzymes involved in α-synuclein degradation and thus a potential therapeutic target [364]. For example, reduced Cathepsin D expression results in α-synuclein accumulation and degeneration of the dopaminergic system in experimental models and in patients with PD [365], whereas overexpression of Cathepsin D reduces the pathology associated with α-synuclein accumulation [366].

In addition to ubiquitination, α-synuclein undergoes posttranslational modification by addition of small ubiquitin-related modifier (SUMO) both in vivo and in vitro. Sumoylation of α-synuclein occurs primarily at Lys96 and Lys102 and results in substoichiometric inhibition of α-synuclein aggregation in vitro in favor of smaller, more soluble amorphous aggregates. Conversely, impairment of sumoylation by mutating Lys96 and Lys102 enhanced aggregation and exacerbated α-synuclein toxicity in cultured cells and in a rat model utilizing rAAV vector-driven expression of wild-type and Lys96/102 mutant α-synuclein [367]. Another study reported that proteasomal inhibition induces the accumulation of sumoylated α-synuclein aggregates in aggresome-like structures in cultured cells, consistent with the detection of sumoylated α-synuclein aggregates in PD and DLB brains and consistent with sumoylation promoting aggregation of other disease-related proteins including ataxin. The authors suggest that sumoylation may enhance aggregation since high molecular weight aggregates were absent in sumoylation-deficient cells [368]. However, it is possible that under normal conditions, sumoylation may promote α-synuclein solubility and clearance and that it may serve as a stress response in the face of accumulated insoluble proteins, perhaps even promoting the formation of nontoxic aggresomes. Distinguishing between these possibilities will be necessary to make informed decisions about therapeutic approaches designed to modulate sumoylation as a PD therapeutic.

8 Summary, Conclusions, Outlook

α-Synuclein is a logical target for a putative disease-modifying PD therapeutic. In its native form, α-synuclein plays an important role at cellular membranes, impacting vesicle trafficking and neurotransmitter release. In its aggregated and modified forms, α-synuclein resists clearance and assumes deleterious new functions, leading to neurodegeneration. α-Synuclein is a nonclassical target for pharmacological intervention due to its dynamic and heterogeneous structure. Thus, conventions regarding stoichiometry, occupancy, and pharmacodynamics may not apply. Initial success has been made in elucidating α-synuclein structure, function, and in neutralizing its deleterious forms, particularly in preclinical models including transgenic mice, gene-transduced rats, and, most recently, human neurons derived from induced pluripotent stem cells of PD patients. Small molecules that engage, redirect, and detoxify misfolded forms show great promise, though their mechanisms of action are complex and not fully understood. Cell-to-cell spreading makes extracellular α-synuclein aggregates an attractive target and may obviate the need for cell penetration of molecules targeting this population. In addition, approaches that enhance α-synuclein clearance are rapidly advancing, including α-synuclein-directed immunotherapy, currently in clinical trials. Nevertheless, critical questions remain. For example, what is the composition of the unfolded ensemble that is capable of forming toxic aggregates and how can structures within this ensemble be safely and selectively redirected to functional or at least nontoxic forms? Moreover, what *are* the "moving targets" that comprise in vivo toxic aggregated species? Is lessening α-synuclein gene dosage or enhancing its clearance in humans safe, and if so how much and at what point in disease progression will these be necessary and sufficient to elicit therapeutic benefit? Further, can noninvasive imaging techniques detect α-synuclein pathology prior to motor dysfunction and herald the need for early intervention? Innovations that successfully address these questions should result in effective α-synuclein-directed PD therapeutics.

Acknowledgments I would like to thank Dr. Kelsey Hanson for reading the manuscript and offering helpful suggestions; Dr. Alan Snow, Dr. Sonal Das, Dr. Manfred Weigele, and Marisa-Claire Yadon for helpful discussions; and the Michael J. Fox Foundation for Parkinson's Disease Research for their support through a LEAPS Award to ProteoTech.

References

1. Spillantini MG, Schmidt ML, Lee VM, Trojanowski JQ, Jakes R, Goedert M (1997) Alpha-synuclein in Lewy bodies. Nature 388:839–840
2. Hawkes CH, Del Tredici K, Braak H (2010) A timeline for Parkinson's disease. Parkinsonism Relat Disord 16:79–84
3. Spillantini MG, Crowther RA, Jakes R, Hasegawa M, Goedert M (1998) Alpha-synuclein in filamentous inclusions of Lewy bodies from Parkinson's disease and dementia with Lewy bodies. Proc Natl Acad Sci U S A 95:6469–6473
4. Baba M, Nakajo S, Tu PH, Tomita T, Nakaya K, Lee VM, Trojanowski JQ, Iwatsubo T (1998) Aggregation of alpha-synuclein in Lewy bodies of sporadic Parkinson's disease and dementia with Lewy bodies. Am J Pathol 152:879–884
5. Galvin JE, Uryu K, Lee VM, Trojanowski JQ (1999) Axon pathology in Parkinson's disease and Lewy body dementia hippocampus contains alpha-, beta-, and gamma-synuclein. Proc Natl Acad Sci U S A 96:13450–13455
6. Conway KA, Harper JD, Lansbury PT (1998) Accelerated in vitro fibril formation by a mutant alpha-synuclein linked to early-onset Parkinson disease. Nat Med 4:1318–1320
7. Narhi L, Wood SJ, Steavenson S, Jiang Y, Wu GM, Anafi D, Kaufman SA, Martin F, Sitney K, Denis P, Louis JC, Wypych J, Biere AL, Citron M (1999) Both familial Parkinson's disease mutations accelerate alpha-synuclein aggregation. J Biol Chem 274:9843–9846

8. Choi W, Zibaee S, Jakes R, Serpell LC, Davletov B, Crowther RA, Goedert M (2004) Mutation E46K increases phospholipid binding and assembly into filaments of human alpha-synuclein. FEBS Lett 576:363–368
9. Polymeropoulos MH, Lavedan C, Leroy E, Ide SE, Dehejia A, Dutra A, Pike B, Root H, Rubenstein J, Boyer R, Stenroos ES, Chandrasekharappa S, Athanassiadou A, Papapetropoulos T, Johnson WG, Lazzarini AM, Duvoisin RC, Di Iorio G, Golbe LI, Nussbaum RL (1997) Mutation in the alpha-synuclein gene identified in families with Parkinson's disease. Science 276:2045–2047
10. Singleton AB, Farrer M, Johnson J, Singleton A, Hague S, Kachergus J, Hulihan M, Peuralinna T, Dutra A, Nussbaum R, Lincoln S, Crawley A, Hanson M, Maraganore D, Adler C, Cookson MR, Muenter M, Baptista M, Miller D, Blancato J, Hardy J, Gwinn-Hardy K (2003) Alpha-synuclein locus triplication causes Parkinson's disease. Science 302:841
11. Ibanez P, Bonnet AM, Debarges B, Lohmann E, Tison F, Pollak P, Agid Y, Durr A, Brice A (2004) Causal relation between alpha-synuclein gene duplication and familial Parkinson's disease. Lancet 364:1169–1171
12. Chartier-Harlin MC, Kachergus J, Roumier C, Mouroux V, Douay X, Lincoln S, Levecque C, Larvor L, Andrieux J, Hulihan M, Waucquier N, Defebvre L, Amouyel P, Farrer M, Destee A (2004) Alpha-synuclein locus duplication as a cause of familial Parkinson's disease. Lancet 364:1167–1169
13. Ikeuchi T, Kakita A, Shiga A, Kasuga K, Kaneko H, Tan CF, Idezuka J, Wakabayashi K, Onodera O, Iwatsubo T, Nishizawa M, Takahashi H, Ishikawa A (2008) Patients homozygous and heterozygous for SNCA duplication in a family with Parkinsonism and dementia. Arch Neurol 65:514–519
14. Simon-Sanchez J, Schulte C, Bras JM, Sharma M, Gibbs JR, Berg D, Paisan-Ruiz C, Lichtner P, Scholz SW, Hernandez DG, Kruger R, Federoff M, Klein C, Goate A, Perlmutter J, Bonin M, Nalls MA, Illig T, Gieger C, Houlden H, Steffens M, Okun MS, Racette BA, Cookson MR, Foote KD, Fernandez HH, Traynor BJ, Schreiber S, Arepalli S, Zonozi R, Gwinn K, van der Brug M, Lopez G, Chanock SJ, Schatzkin A, Park Y, Hollenbeck A, Gao J, Huang X, Wood NW, Lorenz D, Deuschl G, Chen H, Riess O, Hardy JA, Singleton AB, Gasser T (2009) Genome-wide association study reveals genetic risk underlying Parkinson's disease. Nat Genet 41:1308–1312
15. Hattori N, Shimura H, Kubo S, Wang M, Shimizu N, Tanaka K, Mizuno Y (2000) Importance of familial Parkinson's disease and Parkinsonism to the understanding of nigral degeneration in sporadic Parkinson's disease. J Neural Transm Suppl 101–116
16. Lo Bianco C, Schneider BL, Bauer M, Sajadi A, Brice A, Iwatsubo T, Aebischer P (2004) Lentiviral vector delivery of parkin prevents dopaminergic degeneration in an alpha-synuclein rat model of Parkinson's disease. Proc Natl Acad Sci U S A 101:17510–17515
17. Chaudhuri KR, Healy DG, Schapira AH (2006) Non-motor symptoms of Parkinson's disease: diagnosis and management. Lancet Neurol 5:235–245
18. Langston JW (2006) The Parkinson's complex: Parkinsonism is just the tip of the iceberg. Ann Neurol 59:591–596
19. Levitan K, Chereau D, Cohen SI, Knowles TP, Dobson CM, Fink AL, Anderson JP, Goldstein JM, Millhauser GL (2011) Conserved C-terminal charge exerts a profound influence on the aggregation rate of alpha-synuclein. J Mol Biol 411:329–333
20. Li J, Uversky VN, Fink AL (2001) Effect of familial Parkinson's disease point mutations A30P and A53T on the structural properties, aggregation, and fibrillation of human alpha-synuclein. Biochemistry 40:11604–11613
21. Fauvet B, Mbefo MK, Fares MB, Desobry C, Michael S, Ardah MT, Tsika E, Coune P, Prudent M, Lion N, Eliezer D, Moore DJ, Schneider B, Aebischer P, El-Agnaf OM, Masliah E, Lashuel HA (2012) Alpha-synuclein in central nervous system and from erythrocytes, mammalian cells, and Escherichia coli exists predominantly as disordered monomer. J Biol Chem 287:15345–15364

22. Burre J, Vivona S, Diao J, Sharma M, Brunger AT, Sudhof TC (2013) Properties of native brain alpha-synuclein. Nature 498:E4-6; discussion E6-7
23. Bartels T, Choi JG, Selkoe DJ (2011) Alpha-synuclein occurs physiologically as a helically folded tetramer that resists aggregation. Nature 477:107–110
24. Wang W, Perovic I, Chittuluru J, Kaganovich A, Nguyen LT, Liao J, Auclair JR, Johnson D, Landeru A, Simorellis AK, Ju S, Cookson MR, Asturias FJ, Agar JN, Webb BN, Kang C, Ringe D, Petsko GA, Pochapsky TC, Hoang QQ (2011) A soluble alpha-synuclein construct forms a dynamic tetramer. Proc Natl Acad Sci U S A 108:17797–17802
25. Dettmer U, Newman AJ, Luth ES, Bartels T, Selkoe D (2013) In vivo cross-linking reveals principally oligomeric forms of alpha-synuclein and beta-synuclein in neurons and non-neural cells. J Biol Chem 288:6371–6385
26. Gould N, Mor D, Lightfoot R, Malkus K, Giasson B, Ischiropoulos H (2014) Evidence of native alpha-synuclein conformers in the human brain. J Biol Chem 289:7929–7934
27. Wood SJ, Wypych J, Steavenson S, Louis JC, Citron M, Biere AL (1999) Alpha-synuclein fibrillogenesis is nucleation-dependent. Implications for the pathogenesis of Parkinson's disease. J Biol Chem 274:19509–19512
28. Sandal M, Valle F, Tessari I, Mammi S, Bergantino E, Musiani F, Brucale M, Bubacco L, Samori B (2008) Conformational equilibria in monomeric alpha-synuclein at the single-molecule level. PLoS Biol 6:e6
29. Kaylor J, Bodner N, Edridge S, Yamin G, Hong DP, Fink AL (2005) Characterization of oligomeric intermediates in alpha-synuclein fibrillation: FRET studies of Y125W/Y133F/Y136F alpha-synuclein. J Mol Biol 353:357–372
30. Emadi S, Kasturirangan S, Wang MS, Schulz P, Sierks MR (2009) Detecting morphologically distinct oligomeric forms of alpha-synuclein. J Biol Chem 284:11048–11058
31. Conway KA, Harper JD, Lansbury PT Jr (2000) Fibrils formed in vitro from alpha-synuclein and two mutant forms linked to Parkinson's disease are typical amyloid. Biochemistry 39:2552–2563
32. Cremades N, Cohen SI, Deas E, Abramov AY, Chen AY, Orte A, Sandal M, Clarke RW, Dunne P, Aprile FA, Bertoncini CW, Wood NW, Knowles TP, Dobson CM, Klenerman D (2012) Direct observation of the interconversion of normal and toxic forms of alpha-synuclein. Cell 149:1048–1059
33. Maroteaux L, Campanelli JT, Scheller RH (1988) Synuclein: a neuron-specific protein localized to the nucleus and presynaptic nerve terminal. J Neurosci 8:2804–2815
34. George S, Rey NL, Reichenbach N, Steiner JA, Brundin P (2013) Alpha-synuclein: the long distance runner. Brain Pathol 23:350–357
35. Grathwohl SA, Steiner JA, Britschgi M, Brundin P (2013) Mind the gut: secretion of alpha-synuclein by enteric neurons. J Neurochem 125:487–490
36. Schapira AH, Tolosa E (2010) Molecular and clinical prodrome of Parkinson disease: implications for treatment. Nat Rev Neurol 6:309–317
37. Wu KP, Kim S, Fela DA, Baum J (2008) Characterization of conformational and dynamic properties of natively unfolded human and mouse alpha-synuclein ensembles by NMR: implication for aggregation. J Mol Biol 378:1104–1115
38. Allison JR, Varnai P, Dobson CM, Vendruscolo M (2009) Determination of the free energy landscape of alpha-synuclein using spin label nuclear magnetic resonance measurements. J Am Chem Soc 131:18314–18326
39. Haque F, Pandey AP, Cambrea LR, Rochet JC, Hovis JS (2010) Adsorption of alpha-synuclein on lipid bilayers: modulating the structure and stability of protein assemblies. J Phys Chem B 114:4070–4081
40. Smith DP, Tew DJ, Hill AF, Bottomley SP, Masters CL, Barnham KJ, Cappai R (2008) Formation of a high affinity lipid-binding intermediate during the early aggregation phase of alpha-synuclein. Biochemistry 47:1425–1434
41. Dyson HJ, Wright PE (2005) Intrinsically unstructured proteins and their functions. Nat Rev Mol Cell Biol 6:197–208

42. Sung YH, Eliezer D (2007) Residual structure, backbone dynamics, and interactions within the synuclein family. J Mol Biol 372:689–707
43. Kruger R, Kuhn W, Muller T, Woitalla D, Graeber M, Kosel S, Przuntek H, Epplen JT, Schols L, Riess O (1998) Ala30Pro mutation in the gene encoding alpha-synuclein in Parkinson's disease. Nat Genet 18:106–108
44. Zarranz JJ, Alegre J, Gomez-Esteban JC, Lezcano E, Ros R, Ampuero I, Vidal L, Hoenicka J, Rodriguez O, Atares B, Llorens V, Gomez Tortosa E, del Ser T, Munoz DG, de Yebenes JG (2004) The new mutation, E46K, of alpha-synuclein causes Parkinson and Lewy body dementia. Ann Neurol 55:164–173
45. Appel-Cresswell S, Vilarino-Guell C, Encarnacion M, Sherman H, Yu I, Shah B, Weir D, Thompson C, Szu-Tu C, Trinh J, Aasly JO, Rajput A, Rajput AH, Jon Stoessl A, Farrer MJ (2013) Alpha-synuclein p.H50Q, a novel pathogenic mutation for Parkinson's disease. Mov Disord 28:811–813
46. Lesage S, Anheim M, Letournel F, Bousset L, Honore A, Rozas N, Pieri L, Madiona K, Durr A, Melki R, Verny C, Brice A (2013) G51D alpha-synuclein mutation causes a novel Parkinsonian-pyramidal syndrome. Ann Neurol 73:459–471
47. Proukakis C, Dudzik CG, Brier T, MacKay DS, Cooper JM, Millhauser GL, Houlden H, Schapira AH (2013) A novel alpha-synuclein missense mutation in Parkinson disease. Neurology 80:1062–1064
48. Bussell R Jr, Eliezer D (2001) Residual structure and dynamics in Parkinson's disease-associated mutants of alpha-synuclein. J Biol Chem 276:45996–46003
49. Bertoncini CW, Jung YS, Fernandez CO, Hoyer W, Griesinger C, Jovin TM, Zweckstetter M (2005) Release of long-range tertiary interactions potentiates aggregation of natively unstructured alpha-synuclein. Proc Natl Acad Sci U S A 102:1430–1435
50. Dedmon MM, Lindorff-Larsen K, Christodoulou J, Vendruscolo M, Dobson CM (2005) Mapping long-range interactions in alpha-synuclein using spin-label NMR and ensemble molecular dynamics simulations. J Am Chem Soc 127:476–477
51. Hong DP, Xiong W, Chang JY, Jiang C (2011) The role of the C-terminus of human alpha-synuclein: intra-disulfide bonds between the C-terminus and other regions stabilize non-fibrillar monomeric isomers. FEBS Lett 585:561–566
52. Crowther RA, Jakes R, Spillantini MG, Goedert M (1998) Synthetic filaments assembled from C-terminally truncated alpha-synuclein. FEBS Lett 436:309–312
53. Rospigliosi CC, McClendon S, Schmid AW, Ramlall TF, Barre P, Lashuel HA, Eliezer D (2009) E46K Parkinson's-linked mutation enhances C-terminal-to-N-terminal contacts in alpha-synuclein. J Mol Biol 388:1022–1032
54. McClendon S, Rospigliosi CC, Eliezer D (2009) Charge neutralization and collapse of the C-terminal tail of alpha-synuclein at low pH. Protein Sci 18:1531–1540
55. Cho MK, Nodet G, Kim HY, Jensen MR, Bernado P, Fernandez CO, Becker S, Blackledge M, Zweckstetter M (2009) Structural characterization of alpha-synuclein in an aggregation prone state. Protein Sci 18:1840–1846
56. Trexler AJ, Rhoades E (2010) Single molecule characterization of alpha-synuclein in aggregation-prone states. Biophys J 99:3048–3055
57. Ullman O, Fisher CK, Stultz CM (2011) Explaining the structural plasticity of alpha-synuclein. J Am Chem Soc 133:19536–19546
58. Fujiwara H, Hasegawa M, Dohmae N, Kawashima A, Masliah E, Goldberg MS, Shen J, Takio K, Iwatsubo T (2002) Alpha-synuclein is phosphorylated in synucleinopathy lesions. Nat Cell Biol 4:160–164
59. Dikiy I, Eliezer D (2012) Folding and misfolding of alpha-synuclein on membranes. Biochim Biophys Acta 1818:1013–1018
60. Paleologou KE, Oueslati A, Shakked G, Rospigliosi CC, Kim HY, Lamberto GR, Fernandez CO, Schmid A, Chegini F, Gai WP, Chiappe D, Moniatte M, Schneider BL, Aebischer P, Eliezer D, Zweckstetter M, Masliah E, Lashuel HA (2010) Phosphorylation at S87 is

enhanced in synucleinopathies, inhibits alpha-synuclein oligomerization, and influences synuclein-membrane interactions. J Neurosci 30:3184–3198
61. Ahmad B, Chen Y, Lapidus LJ (2012) Aggregation of alpha-synuclein is kinetically controlled by intramolecular diffusion. Proc Natl Acad Sci U S A 109:2336–2341
62. Pandey N, Strider J, Nolan WC, Yan SX, Galvin JE (2008) Curcumin inhibits aggregation of alpha-synuclein. Acta Neuropathol 115:479–489
63. Singh PK, Kotia V, Ghosh D, Mohite GM, Kumar A, Maji SK (2013) Curcumin modulates alpha-synuclein aggregation and toxicity. ACS Chem Neurosci 4:393–407
64. Ahmad B, Lapidus LJ (2012) Curcumin prevents aggregation in alpha-synuclein by increasing reconfiguration rate. J Biol Chem 287:9193–9199
65. Davidson WS, Jonas A, Clayton DF, George JM (1998) Stabilization of alpha-synuclein secondary structure upon binding to synthetic membranes. J Biol Chem 273:9443–9449
66. Jo E, McLaurin J, Yip CM, St George-Hyslop P, Fraser PE (2000) Alpha-synuclein membrane interactions and lipid specificity. J Biol Chem 275:34328–34334
67. Zhu M, Fink AL (2003) Lipid binding inhibits alpha-synuclein fibril formation. J Biol Chem 278:16873–16877
68. Fortin DL, Troyer MD, Nakamura K, Kubo S, Anthony MD, Edwards RH (2004) Lipid rafts mediate the synaptic localization of alpha-synuclein. J Neurosci 24:6715–6723
69. Unni VK, Weissman TA, Rockenstein E, Masliah E, McLean PJ, Hyman BT (2010) In vivo imaging of alpha-synuclein in mouse cortex demonstrates stable expression and differential subcellular compartment mobility. PLoS One 5:e10589
70. Varkey J, Isas JM, Mizuno N, Jensen MB, Bhatia VK, Jao CC, Petrlova J, Voss JC, Stamou DG, Steven AC, Langen R (2010) Membrane curvature induction and tubulation are common features of synucleins and apolipoproteins. J Biol Chem 285:32486–32493
71. Pandey AP, Haque F, Rochet JC, Hovis JS (2011) Alpha-synuclein-induced tubule formation in lipid bilayers. The journal of physical chemistry. B 115:5886–5893
72. Kahle PJ, Neumann M, Ozmen L, Muller V, Jacobsen H, Schindzielorz A, Okochi M, Leimer U, van Der Putten H, Probst A, Kremmer E, Kretzschmar HA, Haass C (2000) Subcellular localization of wild-type and Parkinson's disease-associated mutant alpha-synuclein in human and transgenic mouse brain. J Neurosci 20:6365–6373
73. Fortin DL, Nemani VM, Voglmaier SM, Anthony MD, Ryan TA, Edwards RH (2005) Neural activity controls the synaptic accumulation of alpha-synuclein. J Neurosci 25:10913–10921
74. Leftin A, Job C, Beyer K, Brown MF (2013) Solid-state ^{13}C NMR reveals annealing of raft-like membranes containing cholesterol by the intrinsically disordered protein alpha-synuclein. J Mol Biol 425:2973–2987
75. Murphy DD, Rueter SM, Trojanowski JQ, Lee VM (2000) Synucleins are developmentally expressed, and alpha-synuclein regulates the size of the presynaptic vesicular pool in primary hippocampal neurons. J Neurosci 20:3214–3220
76. Cabin DE, Shimazu K, Murphy D, Cole NB, Gottschalk W, McIlwain KL, Orrison B, Chen A, Ellis CE, Paylor R, Lu B, Nussbaum RL (2002) Synaptic vesicle depletion correlates with attenuated synaptic responses to prolonged repetitive stimulation in mice lacking alpha-synuclein. J Neurosci 22:8797–8807
77. Abeliovich A, Schmitz Y, Farinas I, Choi-Lundberg D, Ho WH, Castillo PE, Shinsky N, Verdugo JM, Armanini M, Ryan A, Hynes M, Phillips H, Sulzer D, Rosenthal A (2000) Mice lacking alpha-synuclein display functional deficits in the nigrostriatal dopamine system. Neuron 25:239–252
78. Yavich L, Tanila H, Vepsalainen S, Jakala P (2004) Role of alpha-synuclein in presynaptic dopamine recruitment. J Neurosci 24:11165–11170
79. Larsen KE, Schmitz Y, Troyer MD, Mosharov E, Dietrich P, Quazi AZ, Savalle M, Nemani V, Chaudhry FA, Edwards RH, Stefanis L, Sulzer D (2006) Alpha-synuclein overexpression in PC12 and chromaffin cells impairs catecholamine release by interfering with a late step in exocytosis. J Neurosci 26:11915–11922

80. Nemani VM, Lu W, Berge V, Nakamura K, Onoa B, Lee MK, Chaudhry FA, Nicoll RA, Edwards RH (2010) Increased expression of alpha-synuclein reduces neurotransmitter release by inhibiting synaptic vesicle reclustering after endocytosis. Neuron 65:66–79
81. Scott D, Roy S (2012) Alpha-synuclein inhibits intersynaptic vesicle mobility and maintains recycling-pool homeostasis. J Neurosci 32:10129–10135
82. Masliah E, Rockenstein E, Veinbergs I, Mallory M, Hashimoto M, Takeda A, Sagara Y, Sisk A, Mucke L (2000) Dopaminergic loss and inclusion body formation in alpha-synuclein mice: implications for neurodegenerative disorders. Science 287:1265–1269
83. Boassa D, Berlanga ML, Yang MA, Terada M, Hu J, Bushong EA, Hwang M, Masliah E, George JM, Ellisman MH (2013) Mapping the subcellular distribution of alpha-synuclein in neurons using genetically encoded probes for correlated light and electron microscopy: implications for Parkinson's disease pathogenesis. J Neurosci 33:2605–2615
84. Burre J, Sharma M, Tsetsenis T, Buchman V, Etherton MR, Sudhof TC (2010) Alpha-synuclein promotes SNARE-complex assembly in vivo and in vitro. Science 329:1663–1667
85. DeWitt DC, Rhoades E (2013) Alpha-synuclein can inhibit SNARE-mediated vesicle fusion through direct interactions with lipid bilayers. Biochemistry 52:2385–2387
86. Chandra S, Gallardo G, Fernandez-Chacon R, Schluter OM, Sudhof TC (2005) Alpha-synuclein cooperates with CSPalpha in preventing neurodegeneration. Cell 123:383–396
87. Dauer W, Kholodilov N, Vila M, Trillat AC, Goodchild R, Larsen KE, Staal R, Tieu K, Schmitz Y, Yuan CA, Rocha M, Jackson-Lewis V, Hersch S, Sulzer D, Przedborski S, Burke R, Hen R (2002) Resistance of alpha-synuclein null mice to the Parkinsonian neurotoxin MPTP. Proc Natl Acad Sci U S A 99:14524–14529
88. Song DD, Shults CW, Sisk A, Rockenstein E, Masliah E (2004) Enhanced substantia nigra mitochondrial pathology in human alpha-synuclein transgenic mice after treatment with MPTP. Exp Neurol 186:158–172
89. Thomas B, Mandir AS, West N, Liu Y, Andrabi SA, Stirling W, Dawson VL, Dawson TM, Lee MK (2011) Resistance to MPTP-neurotoxicity in alpha-synuclein knockout mice is complemented by human alpha-synuclein and associated with increased beta-synuclein and Akt activation. PLoS One 6:e16706
90. Leng Y, Chuang DM (2006) Endogenous alpha-synuclein is induced by valproic acid through histone deacetylase inhibition and participates in neuroprotection against glutamate-induced excitotoxicity. J Neurosci 26:7502–7512
91. Cooper AA, Gitler AD, Cashikar A, Haynes CM, Hill KJ, Bhullar B, Liu K, Xu K, Strathearn KE, Liu F, Cao S, Caldwell KA, Caldwell GA, Marsischky G, Kolodner RD, Labaer J, Rochet JC, Bonini NM, Lindquist S (2006) Alpha-synuclein blocks ER-Golgi traffic and Rab1 rescues neuron loss in Parkinson's models. Science 313:324–328
92. Gitler AD, Bevis BJ, Shorter J, Strathearn KE, Hamamichi S, Su LJ, Caldwell KA, Caldwell GA, Rochet JC, McCaffery JM, Barlowe C, Lindquist S (2008) The Parkinson's disease protein alpha-synuclein disrupts cellular Rab homeostasis. Proc Natl Acad Sci U S A 105:145–150
93. Tardiff DF, Jui NT, Khurana V, Tambe MA, Thompson ML, Chung CY, Kamadurai HB, Kim HT, Lancaster AK, Caldwell KA, Caldwell GA, Rochet JC, Buchwald SL, Lindquist S (2013) Yeast reveal a "druggable" Rsp5/Nedd4 network that ameliorates alpha-synuclein toxicity in neurons. Science 342:979–983
94. Lashuel HA, Overk CR, Oueslati A, Masliah E (2013) The many faces of alpha-synuclein: from structure and toxicity to therapeutic target. Nat Rev Neurosci 14:38–48
95. Apetri MM, Maiti NC, Zagorski MG, Carey PR, Anderson VE (2006) Secondary structure of alpha-synuclein oligomers: characterization by raman and atomic force microscopy. J Mol Biol 355:63–71
96. Celej MS, Sarroukh R, Goormaghtigh E, Fidelio GD, Ruysschaert JM, Raussens V (2012) Toxic prefibrillar alpha-synuclein amyloid oligomers adopt a distinctive antiparallel beta-sheet structure. Biochem J 443:719–726

97. Caughey B, Lansbury PT (2003) Protofibrils, pores, fibrils, and neurodegeneration: separating the responsible protein aggregates from the innocent bystanders. Ann Rev Neurosci 26:267–298
98. Fink AL (2006) The aggregation and fibrillation of alpha-synuclein. Acc Chem Res 39:628–634
99. Tsigelny IF, Bar-On P, Sharikov Y, Crews L, Hashimoto M, Miller MA, Keller SH, Platoshyn O, Yuan JX, Masliah E (2007) Dynamics of alpha-synuclein aggregation and inhibition of pore-like oligomer development by beta-synuclein. FEBS J 274:1862–1877
100. Varkey J, Mizuno N, Hegde BG, Cheng N, Steven AC, Langen R (2013) Alpha-synuclein oligomers with broken helical conformation form lipoprotein nanoparticles. J Biol Chem 288:17620–17630
101. Glabe CG (2008) Structural classification of toxic amyloid oligomers. J Biol Chem 283:29639–29643
102. Winner B, Jappelli R, Maji SK, Desplats PA, Boyer L, Aigner S, Hetzer C, Loher T, Vilar M, Campioni S, Tzitzilonis C, Soragni A, Jessberger S, Mira H, Consiglio A, Pham E, Masliah E, Gage FH, Riek R (2011) In vivo demonstration that alpha-synuclein oligomers are toxic. Proc Natl Acad Sci U S A 108:4194–4199
103. Kim HY, Cho MK, Kumar A, Maier E, Siebenhaar C, Becker S, Fernandez CO, Lashuel HA, Benz R, Lange A, Zweckstetter M (2009) Structural properties of pore-forming oligomers of alpha-synuclein. J Am Chem Soc 131:17482–17489
104. Conway KA, Lee SJ, Rochet JC, Ding TT, Williamson RE, Lansbury PT Jr (2000) Acceleration of oligomerization, not fibrillization, is a shared property of both alpha-synuclein mutations linked to early-onset Parkinson's disease: implications for pathogenesis and therapy. Proc Natl Acad Sci U S A 97:571–576
105. Cappai R, Leck SL, Tew DJ, Williamson NA, Smith DP, Galatis D, Sharples RA, Curtain CC, Ali FE, Cherny RA, Culvenor JG, Bottomley SP, Masters CL, Barnham KJ, Hill AF (2005) Dopamine promotes alpha-synuclein aggregation into SDS-resistant soluble oligomers via a distinct folding pathway. FASEB J 19:1377–1379
106. Hinault MP, Cuendet AF, Mattoo RU, Mensi M, Dietler G, Lashuel HA, Goloubinoff P (2010) Stable alpha-synuclein oligomers strongly inhibit chaperone activity of the Hsp70 system by weak interactions with J-domain co-chaperones. J Biol Chem 285:38173–38182
107. Danzer KM, Haasen D, Karow AR, Moussaud S, Habeck M, Giese A, Kretzschmar H, Hengerer B, Kostka M (2007) Different species of alpha-synuclein oligomers induce calcium influx and seeding. J Neurosci 27:9220–9232
108. Zhou W, Long C, Reaney SH, Di Monte DA, Fink AL, Uversky VN (2010) Methionine oxidation stabilizes non-toxic oligomers of alpha-synuclein through strengthening the auto-inhibitory intra-molecular long-range interactions. Biochim Biophys Acta 1802:322–330
109. Beraud D, Hathaway HA, Trecki J, Chasovskikh S, Johnson DA, Johnson JA, Federoff HJ, Shimoji M, Mhyre TR, Maguire-Zeiss KA (2013) Microglial activation and antioxidant responses induced by the Parkinson's disease protein alpha-synuclein. J Neuroimmune Pharmacol 8:94–117
110. Kim C, Ho DH, Suk JE, You S, Michael S, Kang J, Joong Lee S, Masliah E, Hwang D, Lee HJ, Lee SJ (2013) Neuron-released oligomeric alpha-synuclein is an endogenous agonist of TLR2 for paracrine activation of microglia. Nat Commun 4:1562
111. Volles MJ, Lee SJ, Rochet JC, Shtilerman MD, Ding TT, Kessler JC, Lansbury PT Jr (2001) Vesicle permeabilization by protofibrillar alpha-synuclein: implications for the pathogenesis and treatment of Parkinson's disease. Biochemistry 40:7812–7819
112. Lashuel HA, Petre BM, Wall J, Simon M, Nowak RJ, Walz T, Lansbury PT Jr (2002) Alpha-synuclein, especially the Parkinson's disease-associated mutants, forms pore-like annular and tubular protofibrils. J Mol Biol 322:1089–1102
113. Kostka M, Hogen T, Danzer KM, Levin J, Habeck M, Wirth A, Wagner R, Glabe CG, Finger S, Heinzelmann U, Garidel P, Duan W, Ross CA, Kretzschmar H, Giese A (2008)

Single particle characterization of iron-induced pore-forming alpha-synuclein oligomers. J Biol Chem 283:10992–11003
114. Schmidt F, Levin J, Kamp F, Kretzschmar H, Giese A, Botzel K (2012) Single-channel electrophysiology reveals a distinct and uniform pore complex formed by alpha-synuclein oligomers in lipid membranes. PLoS One 7:e42545
115. Tsigelny IF, Sharikov Y, Wrasidlo W, Gonzalez T, Desplats PA, Crews L, Spencer B, Masliah E (2012) Role of alpha-synuclein penetration into the membrane in the mechanisms of oligomer pore formation. FEBS J 279:1000–1013
116. Zhu M, Rajamani S, Kaylor J, Han S, Zhou F, Fink AL (2004) The flavonoid baicalein inhibits fibrillation of alpha-synuclein and disaggregates existing fibrils. J Biol Chem 279:26846–26857
117. Masuda M, Suzuki N, Taniguchi S, Oikawa T, Nonaka T, Iwatsubo T, Hisanaga S, Goedert M, Hasegawa M (2006) Small molecule inhibitors of alpha-synuclein filament assembly. Biochemistry 45:6085–6094
118. Colla E, Jensen PH, Pletnikova O, Troncoso JC, Glabe C, Lee MK (2012) Accumulation of toxic alpha-synuclein oligomer within endoplasmic reticulum occurs in alpha-synucleinopathy in vivo. J Neurosci 32:3301–3305
119. Colla E, Coune P, Liu Y, Pletnikova O, Troncoso JC, Iwatsubo T, Schneider BL, Lee MK (2012) Endoplasmic reticulum stress is important for the manifestations of alpha-synucleinopathy in vivo. J Neurosci 32:3306–3320
120. Smith WW, Jiang H, Pei Z, Tanaka Y, Morita H, Sawa A, Dawson VL, Dawson TM, Ross CA (2005) Endoplasmic reticulum stress and mitochondrial cell death pathways mediate A53T mutant alpha-synuclein-induced toxicity. Hum Mol Genet 14:3801–3811
121. Martin LJ, Pan Y, Price AC, Sterling W, Copeland NG, Jenkins NA, Price DL, Lee MK (2006) Parkinson's disease alpha-synuclein transgenic mice develop neuronal mitochondrial degeneration and cell death. J Neurosci 26:41–50
122. Roostaee A, Beaudoin S, Staskevicius A, Roucou X (2013) Aggregation and neurotoxicity of recombinant alpha-synuclein aggregates initiated by dimerization. Mol Neurodegener 8:5
123. Karpinar DP, Balija MB, Kugler S, Opazo F, Rezaei-Ghaleh N, Wender N, Kim HY, Taschenberger G, Falkenburger BH, Heise H, Kumar A, Riedel D, Fichtner L, Voigt A, Braus GH, Giller K, Becker S, Herzig A, Baldus M, Jackle H, Eimer S, Schulz JB, Griesinger C, Zweckstetter M (2009) Pre-fibrillar alpha-synuclein variants with impaired beta-structure increase neurotoxicity in Parkinson's disease models. EMBO J 28:3256–3268
124. Choi BK, Choi MG, Kim JY, Yang Y, Lai Y, Kweon DH, Lee NK, Shin YK (2013) Large alpha-synuclein oligomers inhibit neuronal SNARE-mediated vesicle docking. Proc Natl Acad Sci U S A 110:4087–4092
125. Wakabayashi K, Tanji K, Mori F, Takahashi H (2007) The Lewy body in Parkinson's disease: molecules implicated in the formation and degradation of alpha-synuclein aggregates. Neuropathology 27:494–506
126. Serpell LC, Berriman J, Jakes R, Goedert M, Crowther RA (2000) Fiber diffraction of synthetic alpha-synuclein filaments shows amyloid-like cross-beta conformation. Proc Natl Acad Sci U S A 97:4897–4902
127. Vilar M, Chou HT, Luhrs T, Maji SK, Riek-Loher D, Verel R, Manning G, Stahlberg H, Riek R (2008) The fold of alpha-synuclein fibrils. Proc Natl Acad Sci U S A 105:8637–8642
128. Miake H, Mizusawa H, Iwatsubo T, Hasegawa M (2002) Biochemical characterization of the core structure of alpha-synuclein filaments. J Biol Chem 277:19213–19219
129. Volpicelli-Daley LA, Luk KC, Patel TP, Tanik SA, Riddle DM, Stieber A, Meaney DF, Trojanowski JQ, Lee VM (2011) Exogenous alpha-synuclein fibrils induce Lewy body pathology leading to synaptic dysfunction and neuron death. Neuron 72:57–71
130. Giehm L, Svergun DI, Otzen DE, Vestergaard B (2011) Low-resolution structure of a vesicle disrupting alpha-synuclein oligomer that accumulates during fibrillation. Proc Natl Acad Sci U S A 108:3246–3251

131. Horvath I, Weise CF, Andersson EK, Chorell E, Sellstedt M, Bengtsson C, Olofsson A, Hultgren SJ, Chapman M, Wolf-Watz M, Almqvist F, Wittung-Stafshede P (2012) Mechanisms of protein oligomerization: inhibitor of functional amyloids templates alpha-synuclein fibrillation. J Am Chem Soc 134:3439–3444
132. Falsone SF, Meyer NH, Schrank E, Leitinger G, Pham CL, Fodero-Tavoletti MT, Holmberg M, Dulle M, Scicluna B, Gesslbauer B, Ruckert HM, Wagner GE, Merle DA, Nollen EA, Kungl AJ, Hill AF, Cappai R, Zangger K (2012) SERF protein is a direct modifier of amyloid fiber assembly. Cell Rep 2:358–371
133. Obeso JA, Rodriguez-Oroz MC, Goetz CG, Marin C, Kordower JH, Rodriguez M, Hirsch EC, Farrer M, Schapira AH, Halliday G (2010) Missing pieces in the Parkinson's disease puzzle. Nat Med 16:653–661
134. Perrin RJ, Woods WS, Clayton DF, George JM (2001) Exposure to long chain polyunsaturated fatty acids triggers rapid multimerization of synucleins. J Biol Chem 276:41958–41962
135. Lee HJ, Choi C, Lee SJ (2002) Membrane-bound alpha-synuclein has a high aggregation propensity and the ability to seed the aggregation of the cytosolic form. J Biol Chem 277:671–678
136. Necula M, Chirita CN, Kuret J (2003) Rapid anionic micelle-mediated alpha-synuclein fibrillization in vitro. J Biol Chem 278:46674–46680
137. Zhu M, Li J, Fink AL (2003) The association of alpha-synuclein with membranes affects bilayer structure, stability, and fibril formation. J Biol Chem 278:40186–40197
138. Sharon R, Bar-Joseph I, Frosch MP, Walsh DM, Hamilton JA, Selkoe DJ (2003) The formation of highly soluble oligomers of alpha-synuclein is regulated by fatty acids and enhanced in Parkinson's disease. Neuron 37:583–595
139. Bodner CR, Maltsev AS, Dobson CM, Bax A (2010) Differential phospholipid binding of alpha-synuclein variants implicated in Parkinson's disease revealed by solution NMR spectroscopy. Biochemistry 49:862–871
140. Anderson VL, Ramlall TF, Rospigliosi CC, Webb WW, Eliezer D (2010) Identification of a helical intermediate in trifluoroethanol-induced alpha-synuclein aggregation. Proc Natl Acad Sci U S A 107:18850–18855
141. Narayanan V, Scarlata S (2001) Membrane binding and self-association of alpha-synucleins. Biochemistry 40:9927–9934
142. Aisenbrey C, Borowik T, Bystrom R, Bokvist M, Lindstrom F, Misiak H, Sani MA, Grobner G (2008) How is protein aggregation in amyloidogenic diseases modulated by biological membranes? Eur Biophys J 37:247–255
143. Georgieva ER, Ramlall TF, Borbat PP, Freed JH, Eliezer D (2008) Membrane-bound alpha-synuclein forms an extended helix: long-distance pulsed ESR measurements using vesicles, bicelles, and rodlike micelles. J Am Chem Soc 130:12856–12857
144. Georgieva ER, Ramlall TF, Borbat PP, Freed JH, Eliezer D (2010) The lipid-binding domain of wild type and mutant alpha-synuclein: compactness and interconversion between the broken and extended helix forms. J Biol Chem 285:28261–28274
145. Maltsev AS, Ying J, Bax A (2012) Impact of N-terminal acetylation of alpha-synuclein on its random coil and lipid binding properties. Biochemistry 51:5004–5013
146. Hodara R, Norris EH, Giasson BI, Mishizen-Eberz AJ, Lynch DR, Lee VM, Ischiropoulos H (2004) Functional consequences of alpha-synuclein tyrosine nitration: diminished binding to lipid vesicles and increased fibril formation. J Biol Chem 279:47746–47753
147. Schildknecht S, Gerding HR, Karreman C, Drescher M, Lashuel HA, Outeiro TF, Di Monte DA, Leist M (2013) Oxidative and nitrative alpha-synuclein modifications and proteostatic stress: implications for disease mechanisms and interventions in synucleinopathies. J Neurochem 125:491–511
148. Uversky VN, Yamin G, Souillac PO, Goers J, Glaser CB, Fink AL (2002) Methionine oxidation inhibits fibrillation of human alpha-synuclein in vitro. FEBS Lett 517:239–244

149. Maltsev AS, Chen J, Levine RL, Bax A (2013) Site-specific interaction between alpha-synuclein and membranes probed by NMR-observed methionine oxidation rates. J Am Chem Soc 135:2943–2946
150. Wassef R, Haenold R, Hansel A, Brot N, Heinemann SH, Hoshi T (2007) Methionine sulfoxide reductase A and a dietary supplement S-methyl-L-cysteine prevent Parkinson's-like symptoms. J Neurosci 27:12808–12816
151. Liu F, Hindupur J, Nguyen JL, Ruf KJ, Zhu J, Schieler JL, Bonham CC, Wood KV, Davisson VJ, Rochet JC (2008) Methionine sulfoxide reductase A protects dopaminergic cells from Parkinson's disease-related insults. Free Radic Biol Med 45:242–255
152. Galasko DR, Peskind E, Clark CM, Quinn JF, Ringman JM, Jicha GA, Cotman C, Cottrell B, Montine TJ, Thomas RG, Aisen P, Alzheimer's Disease Cooperative Study (2012) Antioxidants for Alzheimer disease: a randomized clinical trial with cerebrospinal fluid biomarker measures. Arch Neurol 69:836–841
153. Hashimoto M, Hsu LJ, Xia Y, Takeda A, Sisk A, Sundsmo M, Masliah E (1999) Oxidative stress induces amyloid-like aggregate formation of NACP/alpha-synuclein in vitro. Neuroreport 10:717–721
154. Ostrerova-Golts N, Petrucelli L, Hardy J, Lee JM, Farer M, Wolozin B (2000) The A53T alpha-synuclein mutation increases iron-dependent aggregation and toxicity. J Neurosci 20:6048–6054
155. Park SS, Schulz EM, Lee D (2007) Disruption of dopamine homeostasis underlies selective neurodegeneration mediated by alpha-synuclein. Eur J Neurosci 26:3104–3112
156. Uversky VN, Li J, Fink AL (2001) Metal-triggered structural transformations, aggregation, and fibrillation of human alpha-synuclein. A possible molecular NK between Parkinson's disease and heavy metal exposure. J Biol Chem 276:44284–44296
157. Yamin G, Glaser CB, Uversky VN, Fink AL (2003) Certain metals trigger fibrillation of methionine-oxidized alpha-synuclein. J Biol Chem 278:27630–27635
158. Gao HM, Kotzbauer PT, Uryu K, Leight S, Trojanowski JQ, Lee VM (2008) Neuroinflammation and oxidation/nitration of alpha-synuclein linked to dopaminergic neurodegeneration. J Neurosci 28:7687–7698
159. Chung CY, Khurana V, Auluck PK, Tardiff DF, Mazzulli JR, Soldner F, Baru V, Lou Y, Freyzon Y, Cho S, Mungenast AE, Muffat J, Mitalipova M, Pluth MD, Jui NT, Schule B, Lippard SJ, Tsai LH, Krainc D, Buchwald SL, Jaenisch R, Lindquist S (2013) Identification and rescue of alpha-synuclein toxicity in Parkinson patient-derived neurons. Science 342:983–987
160. Giasson BI, Duda JE, Murray IV, Chen Q, Souza JM, Hurtig HI, Ischiropoulos H, Trojanowski JQ, Lee VM (2000) Oxidative damage linked to neurodegeneration by selective alpha-synuclein nitration in synucleinopathy lesions. Science 290:985–989
161. Fernandez E, Garcia-Moreno JM, Pablos AM, Chacon J (2013) May the evaluation of nitrosative stress through selective increase of 3-nitrotyrosine proteins other than nitroalbumin and dominant tyrosine-125/136 nitrosylation of serum alpha-synuclein serve for diagnosis of sporadic Parkinson's disease? Antioxid Redox Signal 19:912–918
162. McCormack AL, Mak SK, Di Monte DA (2012) Increased alpha-synuclein phosphorylation and nitration in the aging primate substantia nigra. Cell Death Dis 3:e315
163. McCormack AL, Mak SK, Shenasa M, Langston WJ, Forno LS, Di Monte DA (2008) Pathologic modifications of alpha-synuclein in 1-methyl-4-phenyl-1,2,3,6-tetrahydropyridine (MPTP)-treated squirrel monkeys. J Neuropathol Exp Neurol 67:793–802
164. Hokenson MJ, Uversky VN, Goers J, Yamin G, Munishkina LA, Fink AL (2004) Role of individual methionines in the fibrillation of methionine-oxidized alpha-synuclein. Biochemistry 43:4621–4633
165. Glaser CB, Yamin G, Uversky VN, Fink AL (2005) Methionine oxidation, alpha-synuclein and Parkinson's disease. Biochim Biophys Acta 1703:157–169
166. Yamin G, Uversky VN, Fink AL (2003) Nitration inhibits fibrillation of human alpha-synuclein in vitro by formation of soluble oligomers. FEBS Lett 542:147–152

167. Uversky VN, Yamin G, Munishkina LA, Karymov MA, Millett IS, Doniach S, Lyubchenko YL, Fink AL (2005) Effects of nitration on the structure and aggregation of alpha-synuclein. Brain Res Mol Brain Res 134:84–102
168. Souza JM, Giasson BI, Chen Q, Lee VM, Ischiropoulos H (2000) Dityrosine cross-linking promotes formation of stable alpha-synuclein polymers. Implication of nitrative and oxidative stress in the pathogenesis of neurodegenerative synucleinopathies. J Biol Chem 275:18344–18349
169. Paxinou E, Chen Q, Weisse M, Giasson BI, Norris EH, Rueter SM, Trojanowski JQ, Lee VM, Ischiropoulos H (2001) Induction of alpha-synuclein aggregation by intracellular nitrative insult. J Neurosci 21:8053–8061
170. Qin Z, Hu D, Han S, Reaney SH, Di Monte DA, Fink AL (2007) Effect of 4-hydroxy-2-nonenal modification on alpha-synuclein aggregation. J Biol Chem 282:5862–5870
171. Bae EJ, Ho DH, Park E, Jung JW, Cho K, Hong JH, Lee HJ, Kim KP, Lee SJ (2013) Lipid peroxidation product 4-hydroxy-2-nonenal promotes seeding-capable oligomer formation and cell-to-cell transfer of alpha-synuclein. Antioxid Redox Signal 18:770–783
172. De Franceschi G, Frare E, Pivato M, Relini A, Penco A, Greggio E, Bubacco L, Fontana A, de Laureto PP (2011) Structural and morphological characterization of aggregated species of alpha-synuclein induced by docosahexaenoic acid. J Biol Chem 286:22262–22274
173. Conway KA, Rochet JC, Bieganski RM, Lansbury PT Jr (2001) Kinetic stabilization of the alpha-synuclein protofibril by a dopamine-alpha-synuclein adduct. Science 294:1346–1349
174. Chan T, Chow AM, Cheng XR, Tang DW, Brown IR, Kerman K (2012) Oxidative stress effect of dopamine on alpha-synuclein: electroanalysis of solvent interactions. ACS Chem Neurosci 3:569–574
175. Li HT, Lin DH, Luo XY, Zhang F, Ji LN, Du HN, Song GQ, Hu J, Zhou JW, Hu HY (2005) Inhibition of alpha-synuclein fibrillization by dopamine analogs via reaction with the amino groups of alpha-synuclein. Implication for dopaminergic neurodegeneration. FEBS J 272:3661–3672
176. Nakaso K, Tajima N, Ito S, Teraoka M, Yamashita A, Horikoshi Y, Kikuchi D, Mochida S, Nakashima K, Matsura T (2013) Dopamine-mediated oxidation of methionine 127 in alpha-synuclein causes cytotoxicity and oligomerization of alpha-synuclein. PLoS One 8:e55068
177. Zhou W, Gallagher A, Hong DP, Long C, Fink AL, Uversky VN (2009) At low concentrations, 3,4-dihydroxyphenylacetic acid (DOPAC) binds non-covalently to alpha-synuclein and prevents its fibrillation. J Mol Biol 388:597–610
178. Rekas A, Knott RB, Sokolova A, Barnham KJ, Perez KA, Masters CL, Drew SC, Cappai R, Curtain CC, Pham CL (2010) The structure of dopamine induced alpha-synuclein oligomers. Eur Biophys J 39:1407–1419
179. Norris EH, Giasson BI, Hodara R, Xu S, Trojanowski JQ, Ischiropoulos H, Lee VM (2005) Reversible inhibition of alpha-synuclein fibrillization by dopaminochrome-mediated conformational alterations. J Biol Chem 280:21212–21219
180. Herrera FE, Chesi A, Paleologou KE, Schmid A, Munoz A, Vendruscolo M, Gustincich S, Lashuel HA, Carloni P (2008) Inhibition of alpha-synuclein fibrillization by dopamine is mediated by interactions with five C-terminal residues and with E83 in the NAC region. PLoS One 3:e3394
181. Latawiec D, Herrera F, Bek A, Losasso V, Candotti M, Benetti F, Carlino E, Kranjc A, Lazzarino M, Gustincich S, Carloni P, Legname G (2010) Modulation of alpha-synuclein aggregation by dopamine analogs. PLoS One 5:e9234
182. Anderson JP, Walker DE, Goldstein JM, de Laat R, Banducci K, Caccavello RJ, Barbour R, Huang J, Kling K, Lee M, Diep L, Keim PS, Shen X, Chataway T, Schlossmacher MG, Seubert P, Schenk D, Sinha S, Gai WP, Chilcote TJ (2006) Phosphorylation of Ser-129 is the dominant pathological modification of alpha-synuclein in familial and sporadic Lewy body disease. J Biol Chem 281:29739–29752

183. Zhou J, Broe M, Huang Y, Anderson JP, Gai WP, Milward EA, Porritt M, Howells D, Hughes AJ, Wang X, Halliday GM (2011) Changes in the solubility and phosphorylation of alpha-synuclein over the course of Parkinson's disease. Acta Neuropathol 121:695–704
184. Wang Y, Shi M, Chung KA, Zabetian CP, Leverenz JB, Berg D, Srulijes K, Trojanowski JQ, Lee VM, Siderowf AD, Hurtig H, Litvan I, Schiess MC, Peskind ER, Masuda M, Hasegawa M, Lin X, Pan C, Galasko D, Goldstein DS, Jensen PH, Yang H, Cain KC, Zhang J (2012) Phosphorylated alpha-synuclein in Parkinson's disease. Sci Transl Med 4:121–120
185. Lue LF, Walker DG, Adler CH, Shill H, Tran H, Akiyama H, Sue LI, Caviness J, Sabbagh MN, Beach TG (2012) Biochemical increase in phosphorylated alpha-synuclein precedes histopathology of Lewy-type synucleinopathies. Brain Pathol 22:745–756
186. Foulds PG, Mitchell JD, Parker A, Turner R, Green G, Diggle P, Hasegawa M, Taylor M, Mann D, Allsop D (2011) Phosphorylated alpha-synuclein can be detected in blood plasma and is potentially a useful biomarker for Parkinson's disease. FASEB J 25:4127–4137
187. Paleologou KE, Schmid AW, Rospigliosi CC, Kim HY, Lamberto GR, Fredenburg RA, Lansbury PT Jr, Fernandez CO, Eliezer D, Zweckstetter M, Lashuel HA (2008) Phosphorylation at Ser-129 but not the phosphomimics S129E/D inhibits the fibrillation of alpha-synuclein. J Biol Chem 283:16895–16905
188. Chen L, Feany MB (2005) Alpha-synuclein phosphorylation controls neurotoxicity and inclusion formation in a Drosophila model of Parkinson disease. Nat Neurosci 8:657–663
189. Chau KY, Ching HL, Schapira AH, Cooper JM (2009) Relationship between alpha synuclein phosphorylation, proteasomal inhibition and cell death: relevance to Parkinson's disease pathogenesis. J Neurochem 110:1005–1013
190. Febbraro F, Sahin G, Farran A, Soares S, Jensen PH, Kirik D, Romero-Ramos M (2013) Ser129D mutant alpha-synuclein induces earlier motor dysfunction while S129A results in distinctive pathology in a rat model of Parkinson's disease. Neurobiol Dis 56:47–58
191. Sato H, Arawaka S, Hara S, Fukushima S, Koga K, Koyama S, Kato T (2011) Authentically phosphorylated alpha-synuclein at Ser129 accelerates neurodegeneration in a rat model of familial Parkinson's disease. J Neurosci 31:16884–16894
192. Lee KW, Chen W, Junn E, Im JY, Grosso H, Sonsalla PK, Feng X, Ray N, Fernandez JR, Chao Y, Masliah E, Voronkov M, Braithwaite SP, Stock JB, Mouradian MM (2011) Enhanced phosphatase activity attenuates alpha-synucleinopathy in a mouse model. J Neurosci 31:6963–6971
193. Chau KY, Cooper JM, Schapira AH (2013) Pramipexole reduces phosphorylation of alpha-synuclein at serine-129. J Mol Neurosci 51:573–580
194. Gorbatyuk OS, Li S, Sullivan LF, Chen W, Kondrikova G, Manfredsson FP, Mandel RJ, Muzyczka N (2008) The phosphorylation state of Ser-129 in human alpha-synuclein determines neurodegeneration in a rat model of Parkinson disease. Proc Natl Acad Sci U S A 105:763–768
195. Azeredo da Silveira S, Schneider BL, Cifuentes-Diaz C, Sage D, Abbas-Terki T, Iwatsubo T, Unser M, Aebischer P (2009) Phosphorylation does not prompt, nor prevent, the formation of alpha-synuclein toxic species in a rat model of Parkinson's disease. Hum Mol Genet 18:872–887
196. Kuwahara T, Tonegawa R, Ito G, Mitani S, Iwatsubo T (2012) Phosphorylation of alpha-synuclein protein at Ser-129 reduces neuronal dysfunction by lowering its membrane binding property in *Caenorhabditis elegans*. J Biol Chem 287:7098–7109
197. Waxman EA, Giasson BI (2008) Specificity and regulation of casein kinase-mediated phosphorylation of alpha-synuclein. J Neuropathol Exp Neurol 67:402–416
198. McFarland NR, Fan Z, Xu K, Schwarzschild MA, Feany MB, Hyman BT, McLean PJ (2009) Alpha-synuclein S129 phosphorylation mutants do not alter nigrostriatal toxicity in a rat model of Parkinson disease. J Neuropathol Exp Neurol 68:515–524

199. Schreurs S, Gerard M, Derua R, Waelkens E, Taymans JM, Baekelandt V, Engelborghs Y (2014) In vitro phosphorylation does not influence the aggregation kinetics of WT alpha-synuclein in contrast to its phosphorylation mutants. Int J Mol Sci 15:1040–1067
200. Oueslati A, Paleologou KE, Schneider BL, Aebischer P, Lashuel HA (2012) Mimicking phosphorylation at serine 87 inhibits the aggregation of human alpha-synuclein and protects against its toxicity in a rat model of Parkinson's disease. J Neurosci 32:1536–1544
201. Chen L, Periquet M, Wang X, Negro A, McLean PJ, Hyman BT, Feany MB (2009) Tyrosine and serine phosphorylation of alpha-synuclein have opposing effects on neurotoxicity and soluble oligomer formation. J Clin Invest 119:3257–3265
202. Lim Y, Kehm VM, Lee EB, Soper JH, Li C, Trojanowski JQ, Lee VM (2011) alpha-Syn suppression reverses synaptic and memory defects in a mouse model of dementia with Lewy bodies. J Neurosci 31:10076–10087
203. Kim EJ, Sung JY, Lee HJ, Rhim H, Hasegawa M, Iwatsubo T, Min Do S, Kim J, Paik SR, Chung KC (2006) Dyrk1A phosphorylates alpha-synuclein and enhances intracellular inclusion formation. J Biol Chem 281:33250–33257
204. Okochi M, Walter J, Koyama A, Nakajo S, Baba M, Iwatsubo T, Meijer L, Kahle PJ, Haass C (2000) Constitutive phosphorylation of the Parkinson's disease associated alpha-synuclein. J Biol Chem 275:390–397
205. Lee G, Tanaka M, Park K, Lee SS, Kim YM, Junn E, Lee SH, Mouradian MM (2004) Casein kinase II-mediated phosphorylation regulates alpha-synuclein/synphilin-1 interaction and inclusion body formation. J Biol Chem 279:6834–6839
206. Ishii A, Nonaka T, Taniguchi S, Saito T, Arai T, Mann D, Iwatsubo T, Hisanaga S, Goedert M, Hasegawa M (2007) Casein kinase 2 is the major enzyme in brain that phosphorylates Ser129 of human alpha-synuclein: implication for alpha-synucleinopathies. FEBS Lett 581:4711–4717
207. Inglis KJ, Chereau D, Brigham EF, Chiou SS, Schobel S, Frigon NL, Yu M, Caccavello RJ, Nelson S, Motter R, Wright S, Chian D, Santiago P, Soriano F, Ramos C, Powell K, Goldstein JM, Babcock M, Yednock T, Bard F, Basi GS, Sham H, Chilcote TJ, McConlogue L, Griswold-Prenner I, Anderson JP (2009) Polo-like kinase 2 (PLK2) phosphorylates alpha-synuclein at serine 129 in central nervous system. J Biol Chem 284:2598–2602
208. Mbefo MK, Paleologou KE, Boucharaba A, Oueslati A, Schell H, Fournier M, Olschewski D, Yin G, Zweckstetter M, Masliah E, Kahle PJ, Hirling H, Lashuel HA (2010) Phosphorylation of synucleins by members of the Polo-like kinase family. J Biol Chem 285:2807–2822
209. Oueslati A, Schneider BL, Aebischer P, Lashuel HA (2013) Polo-like kinase 2 regulates selective autophagic alpha-synuclein clearance and suppresses its toxicity in vivo. Proc Natl Acad Sci U S A 110:E3945–E3954
210. Pronin AN, Morris AJ, Surguchov A, Benovic JL (2000) Synucleins are a novel class of substrates for G protein-coupled receptor kinases. J Biol Chem 275:26515–26522
211. Jiang P, Gan M, Ebrahim AS, Castanedes-Casey M, Dickson DW, Yen SH (2013) Adenosine monophosphate-activated protein kinase overactivation leads to accumulation of alpha-synuclein oligomers and decrease of neurites. Neurobiol Aging 34:1504–1515
212. Qing H, Wong W, McGeer EG, McGeer PL (2009) Lrrk2 phosphorylates alpha synuclein at serine 129: Parkinson disease implications. Biochem Biophys Res Commun 387:149–152
213. Lin X, Parisiadou L, Gu XL, Wang L, Shim H, Sun L, Xie C, Long CX, Yang WJ, Ding J, Chen ZZ, Gallant PE, Tao-Cheng JH, Rudow G, Troncoso JC, Liu Z, Li Z, Cai H (2009) Leucine-rich repeat kinase 2 regulates the progression of neuropathology induced by Parkinson's-disease-related mutant alpha-synuclein. Neuron 64:807–827
214. Henderson-Smith A, Chow D, Meechoovet B, Aziz M, Jacobson SA, Shill HA, Sabbagh MN, Caviness JN, Adler CH, Driver-Dunckley ED, Beach TG, Yin H, Dunckley T (2013) SMG1 identified as a regulator of Parkinson's disease-associated alpha-synuclein through siRNA screening. PLoS One 8:e77711

215. Li W, Lesuisse C, Xu Y, Troncoso JC, Price DL, Lee MK (2004) Stabilization of alpha-synuclein protein with aging and familial Parkinson's disease-linked A53T mutation. J Neurosci 24:7400–7409
216. Bennett MC, Bishop JF, Leng Y, Chock PB, Chase TN, Mouradian MM (1999) Degradation of alpha-synuclein by proteasome. J Biol Chem 274:33855–33858
217. Vogiatzi T, Xilouri M, Vekrellis K, Stefanis L (2008) Wild type alpha-synuclein is degraded by chaperone-mediated autophagy and macroautophagy in neuronal cells. J Biol Chem 283:23542–23556
218. Mak SK, McCormack AL, Manning-Bog AB, Cuervo AM, Di Monte DA (2010) Lysosomal degradation of alpha-synuclein in vivo. J Biol Chem 285:13621–13629
219. Kim C, Lee SJ (2008) Controlling the mass action of alpha-synuclein in Parkinson's disease. J Neurochem 107:303–316
220. Webb JL, Ravikumar B, Atkins J, Skepper JN, Rubinsztein DC (2003) Alpha-synuclein is degraded by both autophagy and the proteasome. J Biol Chem 278:25009–25013
221. Lee HJ, Khoshaghideh F, Patel S, Lee SJ (2004) Clearance of alpha-synuclein oligomeric intermediates via the lysosomal degradation pathway. J Neurosci 24:1888–1896
222. Yu WH, Dorado B, Figueroa HY, Wang L, Planel E, Cookson MR, Clark LN, Duff KE (2009) Metabolic activity determines efficacy of macroautophagic clearance of pathological oligomeric alpha-synuclein. Am J Pathol 175:736–747
223. Crews L, Spencer B, Desplats P, Patrick C, Paulino A, Rockenstein E, Hansen L, Adame A, Galasko D, Masliah E (2010) Selective molecular alterations in the autophagy pathway in patients with Lewy body disease and in models of alpha-synucleinopathy. PLoS One 5:e9313
224. Tanik SA, Schultheiss CE, Volpicelli-Daley LA, Brunden KR, Lee VM (2013) Lewy body-like alpha-synuclein aggregates resist degradation and impair macroautophagy. J Biol Chem 288:15194–15210
225. Cuervo AM, Stefanis L, Fredenburg R, Lansbury PT, Sulzer D (2004) Impaired degradation of mutant alpha-synuclein by chaperone-mediated autophagy. Science 305:1292–1295
226. Martinez-Vicente M, Talloczy Z, Kaushik S, Massey AC, Mazzulli J, Mosharov EV, Hodara R, Fredenburg R, Wu DC, Follenzi A, Dauer W, Przedborski S, Ischiropoulos H, Lansbury PT, Sulzer D, Cuervo AM (2008) Dopamine-modified alpha-synuclein blocks chaperone-mediated autophagy. J Clin Invest 118:777–788
227. Yang Q, She H, Gearing M, Colla E, Lee M, Shacka JJ, Mao Z (2009) Regulation of neuronal survival factor MEF2D by chaperone-mediated autophagy. Science 323:124–127
228. Tanaka Y, Engelender S, Igarashi S, Rao RK, Wanner T, Tanzi RE, Sawa A, Dawson VL, Dawson TM, Ross CA (2001) Inducible expression of mutant alpha-synuclein decreases proteasome activity and increases sensitivity to mitochondria-dependent apoptosis. Hum Mol Genet 10:919–926
229. Snyder H, Mensah K, Theisler C, Lee J, Matouschek A, Wolozin B (2003) Aggregated and monomeric alpha-synuclein bind to the S6' proteasomal protein and inhibit proteasomal function. J Biol Chem 278:11753–11759
230. Nonaka T, Hasegawa M (2009) A cellular model to monitor proteasome dysfunction by alpha-synuclein. Biochemistry 48:8014–8022
231. Bousset L, Pieri L, Ruiz-Arlandis G, Gath J, Jensen PH, Habenstein B, Madiona K, Olieric V, Bockmann A, Meier BH, Melki R (2013) Structural and functional characterization of two alpha-synuclein strains. Nat Commun 4:2575
232. Carlson GA, Westaway D, DeArmond SJ, Peterson-Torchia M, Prusiner SB (1989) Primary structure of prion protein may modify scrapie isolate properties. Proc Natl Acad Sci U S A 86:7475–7479
233. Telling GC, Scott M, Hsiao KK, Foster D, Yang SL, Torchia M, Sidle KC, Collinge J, DeArmond SJ, Prusiner SB (1994) Transmission of Creutzfeldt-Jakob disease from humans to transgenic mice expressing chimeric human-mouse prion protein. Proc Natl Acad Sci U S A 91:9936–9940

234. Guo JL, Covell DJ, Daniels JP, Iba M, Stieber A, Zhang B, Riddle DM, Kwong LK, Xu Y, Trojanowski JQ, Lee VM (2013) Distinct alpha-synuclein strains differentially promote tau inclusions in neurons. Cell 154:103–117
235. Dunning CJ, George S, Brundin P (2013) What's to like about the prion-like hypothesis for the spreading of aggregated alpha-synuclein in Parkinson disease? Prion 7:92–97
236. Watts JC, Giles K, Oehler A, Middleton L, Dexter D, Gentlman S, DeArmond SJ, Prusiner SB (2013) Transmission of multiple systems atrophy prions to transgenic mice. Proc Natl Acad Sci U S A. doi:10.1073/pnas.1318268110
237. Rey NL, Petit GH, Bousset L, Melki R, Brundin P (2013) Transfer of human alpha-synuclein from the olfactory bulb to interconnected brain regions in mice. Acta Neuropathol 126:555–573
238. Kordower JH, Chu Y, Hauser RA, Freeman TB, Olanow CW (2008) Lewy body-like pathology in long-term embryonic nigral transplants in Parkinson's disease. Nat Med 14:504–506
239. Kordower JH, Chu Y, Hauser RA, Olanow CW, Freeman TB (2008) Transplanted dopaminergic neurons develop PD pathologic changes: a second case report. Mov Disord 23:2303–2306
240. Li JY, Englund E, Holton JL, Soulet D, Hagell P, Lees AJ, Lashley T, Quinn NP, Rehncrona S, Bjorklund A, Widner H, Revesz T, Lindvall O, Brundin P (2008) Lewy bodies in grafted neurons in subjects with Parkinson's disease suggest host-to-graft disease propagation. Nat Med 14:501–503
241. Desplats P, Lee HJ, Bae EJ, Patrick C, Rockenstein E, Crews L, Spencer B, Masliah E, Lee SJ (2009) Inclusion formation and neuronal cell death through neuron-to-neuron transmission of alpha-synuclein. Proc Natl Acad Sci U S A 106:13010–13015
242. Luk KC, Song C, O'Brien P, Stieber A, Branch JR, Brunden KR, Trojanowski JQ, Lee VM (2009) Exogenous alpha-synuclein fibrils seed the formation of Lewy body-like intracellular inclusions in cultured cells. Proc Natl Acad Sci U S A 106:20051–20056
243. Lee HJ, Suk JE, Patrick C, Bae EJ, Cho JH, Rho S, Hwang D, Masliah E, Lee SJ (2010) Direct transfer of alpha-synuclein from neuron to astroglia causes inflammatory responses in synucleinopathies. J Biol Chem 285:9262–9272
244. Hansen C, Angot E, Bergstrom AL, Steiner JA, Pieri L, Paul G, Outeiro TF, Melki R, Kallunki P, Fog K, Li JY, Brundin P (2011) Alpha-synuclein propagates from mouse brain to grafted dopaminergic neurons and seeds aggregation in cultured human cells. J Clin Invest 121:715–725
245. Luk KC, Kehm V, Carroll J, Zhang B, O'Brien P, Trojanowski JQ, Lee VM (2012) Pathological alpha-synuclein transmission initiates Parkinson-like neurodegeneration in nontransgenic mice. Science 338:949–953
246. Ulusoy A, Rusconi R, Perez-Revuelta BI, Musgrove RE, Helwig M, Winzen-Reichert B, Di Monte DA (2013) Caudo-rostral brain spreading of alpha-synuclein through vagal connections. EMBO Mol Med 5:1051–1059
247. Lee HJ, Patel S, Lee SJ (2005) Intravesicular localization and exocytosis of alpha-synuclein and its aggregates. J Neurosci 25:6016–6024
248. Bae EJ, Lee HJ, Rockenstein E, Ho DH, Park EB, Yang NY, Desplats P, Masliah E, Lee SJ (2012) Antibody-aided clearance of extracellular alpha-synuclein prevents cell-to-cell aggregate transmission. J Neurosci 32:13454–13469
249. Junn E, Lee KW, Jeong BS, Chan TW, Im JY, Mouradian MM (2009) Repression of alpha-synuclein expression and toxicity by microRNA-7. Proc Natl Acad Sci U S A 106:13052–13057
250. McCormack AL, Mak SK, Henderson JM, Bumcrot D, Farrer MJ, Di Monte DA (2010) Alpha-synuclein suppression by targeted small interfering RNA in the primate substantia nigra. PLoS One 5:e12122

251. Hayashita-Kinoh H, Yamada M, Yokota T, Mizuno Y, Mochizuki H (2006) Down-regulation of alpha-synuclein expression can rescue dopaminergic cells from cell death in the substantia nigra of Parkinson's disease rat model. Biochem Biophys Res Commun 341:1088–1095
252. Outeiro TF, Kontopoulos E, Altmann SM, Kufareva I, Strathearn KE, Amore AM, Volk CB, Maxwell MM, Rochet JC, McLean PJ, Young AB, Abagyan R, Feany MB, Hyman BT, Kazantsev AG (2007) Sirtuin 2 inhibitors rescue alpha-synuclein-mediated toxicity in models of Parkinson's disease. Science 317:516–519
253. Bieschke J, Russ J, Friedrich RP, Ehrnhoefer DE, Wobst H, Neugebauer K, Wanker EE (2010) EGCG remodels mature alpha-synuclein and amyloid-beta fibrils and reduces cellular toxicity. Proc Natl Acad Sci U S A 107:7710–7715
254. Ehrnhoefer DE, Bieschke J, Boeddrich A, Herbst M, Masino L, Lurz R, Engemann S, Pastore A, Wanker EE (2008) EGCG redirects amyloidogenic polypeptides into unstructured, off-pathway oligomers. Nat Struct Mol Biol 15:558–566
255. Grelle G, Otto A, Lorenz M, Frank RF, Wanker EE, Bieschke J (2011) Black tea theaflavins inhibit formation of toxic amyloid-beta and alpha-synuclein fibrils. Biochemistry 50:10624–10636
256. Camilleri A, Zarb C, Caruana M, Ostermeier U, Ghio S, Hogen T, Schmidt F, Giese A, Vassallo N (2013) Mitochondrial membrane permeabilisation by amyloid aggregates and protection by polyphenols. Biochim Biophys Acta 1828:2532–2543
257. Caruana M, Neuner J, Hogen T, Schmidt F, Kamp F, Scerri C, Giese A, Vassallo N (2012) Polyphenolic compounds are novel protective agents against lipid membrane damage by alpha-synuclein aggregates in vitro. Biochim Biophys Acta 1818:2502–2510
258. Youdim KA, Qaiser MZ, Begley DJ, Rice-Evans CA, Abbott NJ (2004) Flavonoid permeability across an in situ model of the blood–brain barrier. Free Radic Biol Med 36:592–604
259. Manach C, Williamson G, Morand C, Scalbert A, Remesy C (2005) Bioavailability and bioefficacy of polyphenols in humans. I. Review of 97 bioavailability studies. Am J Clin Nutr 81:230S–242S
260. Wright B (2013) Forging a modern generation of polyphenol-based therapeutics. Br J Pharmacol 169:844–847
261. Meng X, Munishkina LA, Fink AL, Uversky VN (2009) Molecular mechanisms underlying the flavonoid-induced inhibition of alpha-synuclein fibrillation. Biochemistry 48:8206–8224
262. Hong DP, Fink AL, Uversky VN (2008) Structural characteristics of alpha-synuclein oligomers stabilized by the flavonoid baicalein. J Mol Biol 383:214–223
263. Jiang M, Porat-Shliom Y, Pei Z, Cheng Y, Xiang L, Sommers K, Li Q, Gillardon F, Hengerer B, Berlinicke C, Smith WW, Zack DJ, Poirier MA, Ross CA, Duan W (2010) Baicalein reduces E46K alpha-synuclein aggregation in vitro and protects cells against E46K alpha-synuclein toxicity in cell models of familiar Parkinsonism. J Neurochem 114:419–429
264. Lu JH, Ardah MT, Durairajan SS, Liu LF, Xie LX, Fong WF, Hasan MY, Huang JD, El-Agnaf OM, Li M (2011) Baicalein inhibits formation of alpha-synuclein oligomers within living cells and prevents Abeta peptide fibrillation and oligomerisation. Chembiochem 12:615–624
265. Meng X, Munishkina LA, Fink AL, Uversky VN (2010) Effects of various flavonoids on the alpha-synuclein fibrillation process. Parkinsons Dis 2010:650794
266. Li J, Zhu M, Rajamani S, Uversky VN, Fink AL (2004) Rifampicin inhibits alpha-synuclein fibrillation and disaggregates fibrils. Chem Biol 11:1513–1521
267. Yoshida W, Kobayashi N, Sasaki Y, Ikebukuro K, Sode K (2013) Partial peptide of alpha-synuclein modified with small-molecule inhibitors specifically inhibits amyloid fibrillation of alpha-synuclein. Int J Mol Sci 14:2590–2600
268. Segers-Nolten IM, Wilhelmus MM, Veldhuis G, van Rooijen BD, Drukarch B, Subramaniam V (2008) Tissue transglutaminase modulates alpha-synuclein oligomerization. Protein Sci 17:1395–1402
269. Yamaguchi Y, Masuda M, Sasakawa H, Nonaka T, Hanashima S, Hisanaga S, Kato K, Hasegawa M (2010) Characterization of inhibitor-bound alpha-synuclein dimer: role of

alpha-synuclein N-terminal region in dimerization and inhibitor binding. J Mol Biol 395:445–456
270. Landau M, Sawaya MR, Faull KF, Laganowsky A, Jiang L, Sievers SA, Liu J, Barrio JR, Eisenberg D (2011) Towards a pharmacophore for amyloid. PLoS Biol 9:e1001080
271. Lamberto GR, Binolfi A, Orcellet ML, Bertoncini CW, Zweckstetter M, Griesinger C, Fernandez CO (2009) Structural and mechanistic basis behind the inhibitory interaction of PcTS on alpha-synuclein amyloid fibril formation. Proc Natl Acad Sci U S A 106:21057–21062
272. Porat Y, Abramowitz A, Gazit E (2006) Inhibition of amyloid fibril formation by polyphenols: structural similarity and aromatic interactions as a common inhibition mechanism. Chem Biol Drug Des 67:27–37
273. Rao JN, Dua V, Ulmer TS (2008) Characterization of alpha-synuclein interactions with selected aggregation-inhibiting small molecules. Biochemistry 47:4651–4656
274. Ono K, Hirohata M, Yamada M (2007) Anti-fibrillogenic and fibril-destabilizing activities of anti-Parkinsonian agents for alpha-synuclein fibrils in vitro. J Neurosci Res 85:1547–1557
275. Braga CA, Follmer C, Palhano FL, Khattar E, Freitas MS, Romao L, Di Giovanni S, Lashuel HA, Silva JL, Foguel D (2011) The anti-Parkinsonian drug selegiline delays the nucleation phase of alpha-synuclein aggregation leading to the formation of nontoxic species. J Mol Biol 405:254–273
276. Falsone SF, Leitinger G, Karner A, Kungl AJ, Kosol S, Cappai R, Zangger K (2011) The neurotransmitter serotonin interrupts alpha-synuclein amyloid maturation. Biochim Biophys Acta 1814:553–561
277. Luk KC, Hyde EG, Trojanowski JQ, Lee VM (2007) Sensitive fluorescence polarization technique for rapid screening of alpha-synuclein oligomerization/fibrillization inhibitors. Biochemistry 46:12522–12529
278. Lange KW, Rausch WD, Gsell W, Naumann M, Oestreicher E, Riederer P (1994) Neuroprotection by dopamine agonists. J Neural Transm Suppl 43:183–201
279. Palhagen S, Heinonen E, Hagglund J, Kaugesaar T, Maki-Ikola O, Palm R, Swedish Parkinson Study (2006) Selegiline slows the progression of the symptoms of Parkinson disease. Neurology 66:1200–1206
280. Di Giovanni S, Eleuteri S, Paleologou KE, Yin G, Zweckstetter M, Carrupt PA, Lashuel HA (2010) Entacapone and tolcapone, two catechol O-methyltransferase inhibitors, block fibril formation of alpha-synuclein and beta-amyloid and protect against amyloid-induced toxicity. J Biol Chem 285:14941–14954
281. Prabhudesai S, Sinha S, Attar A, Kotagiri A, Fitzmaurice AG, Lakshmanan R, Ivanova MI, Loo JA, Klarner FG, Schrader T, Stahl M, Bitan G, Bronstein JM (2012) A novel "Molecular Tweezer" inhibitor of alpha-synuclein neurotoxicity in vitro and in vivo. Neurotherapeutics 9:464–476
282. Acharya S, Safaie BM, Wongkongkathep P, Ivanova MI, Attar A, Klarner FG, Schrader T, Loo JA, Bitan G, Lapidus LJ (2014) Molecular basis for preventing alpha-synuclein aggregation by a molecular tweezer. J Biol Chem 289:10727–10737
283. Sun M, Su X, Ding B, He X, Liu X, Yu A, Lou H, Zhai G (2012) Advances in nanotechnology-based delivery systems for curcumin. Nanomedicine (Lond) 7:1085–1100
284. Mythri RB, Harish G, Dubey SK, Misra K, Bharath MM (2011) Glutamoyl diester of the dietary polyphenol curcumin offers improved protection against peroxynitrite-mediated nitrosative stress and damage of brain mitochondria in vitro: implications for Parkinson's disease. Mol Cell Biochem 347:135–143
285. Wagner J, Ryazanov S, Leonov A, Levin J, Shi S, Schmidt F, Prix C, Pan-Montojo F, Bertsch U, Mitteregger-Kretzschmar G, Geissen M, Eiden M, Leidel F, Hirschberger T, Deeg AA, Krauth JJ, Zinth W, Tavan P, Pilger J, Zweckstetter M, Frank T, Bahr M, Weishaupt JH, Uhr M, Urlaub H, Teichmann U, Samwer M, Botzel K, Groschup M, Kretzschmar H, Griesinger C, Giese A (2013) Anle138b: a novel oligomer modulator for

disease-modifying therapy of neurodegenerative diseases such as prion and Parkinson's disease. Acta Neuropathol 125:795–813
286. Shaltiel-Karyo R, Frenkel-Pinter M, Rockenstein E, Patrick C, Levy-Sakin M, Schiller A, Egoz-Matia N, Masliah E, Segal D, Gazit E (2013) A blood–brain barrier (BBB) disrupter is also a potent alpha-synuclein (alpha-syn) aggregation inhibitor: a novel dual mechanism of mannitol for the treatment of Parkinson disease (PD). J Biol Chem 288:17579–17588
287. Rockenstein E, Mallory M, Hashimoto M, Song D, Shults CW, Lang I, Masliah E (2002) Differential neuropathological alterations in transgenic mice expressing alpha-synuclein from the platelet-derived growth factor and Thy-1 promoters. J Neurosci Res 68:568–578
288. Faria C, Jorge CD, Borges N, Tenreiro S, Outeiro TF, Santos H (2013) Inhibition of formation of alpha-synuclein inclusions by mannosylglycerate in a yeast model of Parkinson's disease. Biochim Biophys Acta 1830:4065–4072
289. Tardiff DF, Tucci ML, Caldwell KA, Caldwell GA, Lindquist S (2012) Different 8-hydroxyquinolines protect models of TDP-43 protein, alpha-synuclein, and polyglutamine proteotoxicity through distinct mechanisms. J Biol Chem 287:4107–4120
290. Teraoka M, Nakaso K, Kusumoto C, Katano S, Tajima N, Yamashita A, Zushi T, Ito S, Matsura T (2012) Cytoprotective effect of chlorogenic acid against alpha-synuclein-related toxicity in catecholaminergic PC12 cells. J Clin Biochem Nutr 51:122–127
291. El-Agnaf OM, Paleologou KE, Greer B, Abogrein AM, King JE, Salem SA, Fullwood NJ, Benson FE, Hewitt R, Ford KJ, Martin FL, Harriott P, Cookson MR, Allsop D (2004) A strategy for designing inhibitors of alpha-synuclein aggregation and toxicity as a novel treatment for Parkinson's disease and related disorders. FASEB J 18:1315–1317
292. Huggins KN, Bisaglia M, Bubacco L, Tatarek-Nossol M, Kapurniotu A, Andersen NH (2011) Designed hairpin peptides interfere with amyloidogenesis pathways: fibril formation and cytotoxicity inhibition, interception of the preamyloid state. Biochemistry 50:8202–8212
293. Hashimoto M, Rockenstein E, Mante M, Mallory M, Masliah E (2001) beta-Synuclein inhibits alpha-synuclein aggregation: a possible role as an anti-Parkinsonian factor. Neuron 32:213–223
294. Beyer K, Ariza A (2008) The therapeutical potential of alpha-synuclein antiaggregatory agents for dementia with Lewy bodies. Curr Med Chem 15:2748–2759
295. Uversky VN, Li J, Souillac P, Millett IS, Doniach S, Jakes R, Goedert M, Fink AL (2002) Biophysical properties of the synucleins and their propensities to fibrillate: inhibition of alpha-synuclein assembly by beta- and gamma-synucleins. J Biol Chem 277:11970–11978
296. Masliah E, Tsigelny IF, Wrasidlo W, Rockenstein E (2013) Compounds for inhibiting protein aggregation, and methods for making and using them. United States Patent US 8,450,481 B2
297. Valera E, Masliah E (2013) Immunotherapy for neurodegenerative diseases: focus on alpha-synucleinopathies. Pharmacol Ther 138:311–322
298. Masliah E, Rockenstein E, Adame A, Alford M, Crews L, Hashimoto M, Seubert P, Lee M, Goldstein J, Chilcote T, Games D, Schenk D (2005) Effects of alpha-synuclein immunization in a mouse model of Parkinson's disease. Neuron 46:857–868
299. Nasstrom T, Goncalves S, Sahlin C, Nordstrom E, Screpanti Sundquist V, Lannfelt L, Bergstrom J, Outeiro TF, Ingelsson M (2011) Antibodies against alpha-synuclein reduce oligomerization in living cells. PLoS One 6:e27230
300. Masliah E, Rockenstein E, Mante M, Crews L, Spencer B, Adame A, Patrick C, Trejo M, Ubhi K, Rohn TT, Mueller-Steiner S, Seubert P, Barbour R, McConlogue L, Buttini M, Games D, Schenk D (2011) Passive immunization reduces behavioral and neuropathological deficits in an alpha-synuclein transgenic model of Lewy body disease. PLoS One 6:e19338
301. Schneeberger A, Mandler M, Mattner F, Schmidt W (2012) Vaccination for Parkinson's disease. Parkinsonism Relat Disord 18(Suppl 1):S11–S13
302. Emadi S, Barkhordarian H, Wang MS, Schulz P, Sierks MR (2007) Isolation of a human single chain antibody fragment against oligomeric alpha-synuclein that inhibits aggregation and prevents alpha-synuclein-induced toxicity. J Mol Biol 368:1132–1144

303. Zhou C, Emadi S, Sierks MR, Messer A (2004) A human single-chain Fv intrabody blocks aberrant cellular effects of overexpressed alpha-synuclein. Mol Ther 10:1023–1031
304. Sancenon V, Lee SA, Patrick C, Griffith J, Paulino A, Outeiro TF, Reggiori F, Masliah E, Muchowski PJ (2012) Suppression of alpha-synuclein toxicity and vesicle trafficking defects by phosphorylation at S129 in yeast depends on genetic context. Hum Mol Genet 21:2432–2449
305. Toth G, Gardai SJ, Zago W, Bertoncini CW, Cremades N, Roy SL, Tambe MA, Rochet JC, Galvagnion C, Skibinski G, Finkbeiner S, Bova M, Regnstrom K, Chiou SS, Johnston J, Callaway K, Anderson JP, Jobling MF, Buell AK, Yednock TA, Knowles TP, Vendruscolo M, Christodoulou J, Dobson CM, Schenk D, McConlogue L (2014) Targeting the intrinsically disordered structural ensemble of alpha-synuclein by small molecules as a potential therapeutic strategy for Parkinson's disease. PLoS One 9:e87133
306. Kontopoulos E, Parvin JD, Feany MB (2006) Alpha-synuclein acts in the nucleus to inhibit histone acetylation and promote neurotoxicity. Hum Mol Genet 15:3012–3023
307. Monti B, Gatta V, Piretti F, Raffaelli SS, Virgili M, Contestabile A (2010) Valproic acid is neuroprotective in the rotenone rat model of Parkinson's disease: involvement of alpha-synuclein. Neurotox Res 17:130–141
308. Desplats P, Spencer B, Coffee E, Patel P, Michael S, Patrick C, Adame A, Rockenstein E, Masliah E (2011) Alpha-synuclein sequesters Dnmtl from the nucleus: a novel mechanism for epigenetic alterations in Lewy body diseases. J Biol Chem 286:9031–9037
309. Jowaed A, Schmitt I, Kaut O, Wullner U (2010) Methylation regulates alpha-synuclein expression and is decreased in Parkinson's disease patients' brains. J Neurosci 30:6355–6359
310. Matsumoto L, Takuma H, Tamaoka A, Kurisaki H, Date H, Tsuji S, Iwata A (2010) CpG demethylation enhances alpha-synuclein expression and affects the pathogenesis of Parkinson's disease. PLoS One 5:e15522
311. Schapira AH, Cooper JM, Dexter D, Clark JB, Jenner P, Marsden CD (1990) Mitochondrial complex I deficiency in Parkinson's disease. J Neurochem 54:823–827
312. Kamp F, Exner N, Lutz AK, Wender N, Hegermann J, Brunner B, Nuscher B, Bartels T, Giese A, Beyer K, Eimer S, Winklhofer KF, Haass C (2010) Inhibition of mitochondrial fusion by alpha-synuclein is rescued by PINK1, Parkin and DJ-1. EMBO J 29:3571–3589
313. Vives-Bauza C, Przedborski S (2011) Mitophagy: the latest problem for Parkinson's disease. Trends Mol Med 17:158–165
314. Cali T, Ottolini D, Negro A, Brini M (2012) Alpha-synuclein controls mitochondrial calcium homeostasis by enhancing endoplasmic reticulum-mitochondria interactions. J Biol Chem 287:17914–17929
315. Stichel CC, Zhu XR, Bader V, Linnartz B, Schmidt S, Lubbert H (2007) Mono- and double-mutant mouse models of Parkinson's disease display severe mitochondrial damage. Hum Mol Genet 16:2377–2393
316. Devi L, Raghavendran V, Prabhu BM, Avadhani NG, Anandatheerthavarada HK (2008) Mitochondrial import and accumulation of alpha-synuclein impair complex I in human dopaminergic neuronal cultures and Parkinson disease brain. J Biol Chem 283:9089–9100
317. Banerjee K, Sinha M, Pham Cle L, Jana S, Chanda D, Cappai R, Chakrabarti S (2010) Alpha-synuclein induced membrane depolarization and loss of phosphorylation capacity of isolated rat brain mitochondria: implications in Parkinson's disease. FEBS Lett 584:1571–1576
318. Martin LJ, Semenkow S, Hanaford A, Wong M (2013) The mitochondrial permeability transition pore regulates Parkinson's disease development in mutant alpha-synuclein transgenic mice. Neurobiol Aging 35:1132–1152
319. Li WW, Yang R, Guo JC, Ren HM, Zha XL, Cheng JS, Cai DF (2007) Localization of alpha-synuclein to mitochondria within midbrain of mice. Neuroreport 18:1543–1546
320. Parihar MS, Parihar A, Fujita M, Hashimoto M, Ghafourifar P (2008) Mitochondrial association of alpha-synuclein causes oxidative stress. Cell Mol Life Sci 65:1272–1284

321. Nakamura K, Nemani VM, Wallender EK, Kaehlcke K, Ott M, Edwards RH (2008) Optical reporters for the conformation of alpha-synuclein reveal a specific interaction with mitochondria. J Neurosci 28:12305–12317
322. Hsu LJ, Sagara Y, Arroyo A, Rockenstein E, Sisk A, Mallory M, Wong J, Takenouchi T, Hashimoto M, Masliah E (2000) Alpha-synuclein promotes mitochondrial deficit and oxidative stress. Am J Pathol 157:401–410
323. Buttner S, Bitto A, Ring J, Augsten M, Zabrocki P, Eisenberg T, Jungwirth H, Hutter S, Carmona-Gutierrez D, Kroemer G, Winderickx J, Madeo F (2008) Functional mitochondria are required for alpha-synuclein toxicity in aging yeast. J Biol Chem 283:7554–7560
324. Dryanovski DI, Guzman JN, Xie Z, Galteri DJ, Volpicelli-Daley LA, Lee VM, Miller RJ, Schumacker PT, Surmeier DJ (2013) Calcium entry and alpha-synuclein inclusions elevate dendritic mitochondrial oxidant stress in dopaminergic neurons. J Neurosci 33:10154–10164
325. Buttner S, Habernig L, Broeskamp F, Ruli D, Vogtle FN, Vlachos M, Macchi F, Kuttner V, Carmona-Gutierrez D, Eisenberg T, Ring J, Markaki M, Taskin AA, Benke S, Ruckenstuhl C, Braun R, Van den Haute C, Bammens T, van der Perren A, Frohlich KU, Winderickx J, Kroemer G, Baekelandt V, Tavernarakis N, Kovacs GG, Dengjel J, Meisinger C, Sigrist SJ, Madeo F (2013) Endonuclease G mediates alpha-synuclein cytotoxicity during Parkinson's disease. EMBO J 32:3041–3054
326. Liu G, Zhang C, Yin J, Li X, Cheng F, Li Y, Yang H, Ueda K, Chan P, Yu S (2009) Alpha-synuclein is differentially expressed in mitochondria from different rat brain regions and dose-dependently down-regulates complex I activity. Neurosci Lett 454:187–192
327. Choubey V, Safiulina D, Vaarmann A, Cagalinec M, Wareski P, Kuum M, Zharkovsky A, Kaasik A (2011) Mutant A53T alpha-synuclein induces neuronal death by increasing mitochondrial autophagy. J Biol Chem 286:10814–10824
328. Nakamura K, Nemani VM, Azarbal F, Skibinski G, Levy JM, Egami K, Munishkina L, Zhang J, Gardner B, Wakabayashi J, Sesaki H, Cheng Y, Finkbeiner S, Nussbaum RL, Masliah E, Edwards RH (2011) Direct membrane association drives mitochondrial fission by the Parkinson disease-associated protein alpha-synuclein. J Biol Chem 286:20710–20726
329. Guardia-Laguarta C, Area-Gomez E, Rub C, Liu Y, Magrane J, Becker D, Voos W, Schon EA, Przedborski S (2014) Alpha-synuclein is localized to mitochondria-associated ER membranes. J Neurosci 34:249–259
330. Luk KC, Mills IP, Trojanowski JQ, Lee VM (2008) Interactions between Hsp70 and the hydrophobic core of alpha-synuclein inhibit fibril assembly. Biochemistry 47:12614–12625
331. Dedmon MM, Christodoulou J, Wilson MR, Dobson CM (2005) Heat shock protein 70 inhibits alpha-synuclein fibril formation via preferential binding to prefibrillar species. J Biol Chem 280:14733–14740
332. Klucken J, Shin Y, Masliah E, Hyman BT, McLean PJ (2004) Hsp70 reduces alpha-synuclein aggregation and toxicity. J Biol Chem 279:25497–25502
333. Auluck PK, Chan HY, Trojanowski JQ, Lee VM, Bonini NM (2002) Chaperone suppression of alpha-synuclein toxicity in a Drosophila model for Parkinson's disease. Science 295:865–868
334. Danzer KM, Ruf WP, Putcha P, Joyner D, Hashimoto T, Glabe C, Hyman BT, McLean PJ (2011) Heat-shock protein 70 modulates toxic extracellular alpha-synuclein oligomers and rescues trans-synaptic toxicity. FASEB J 25:326–336
335. Kilpatrick K, Novoa JA, Hancock T, Guerriero CJ, Wipf P, Brodsky JL, Segatori L (2013) Chemical induction of Hsp70 reduces alpha-synuclein aggregation in neuroglioma cells. ACS Chem Biol 8:1460–1468
336. Lo Bianco C, Shorter J, Regulier E, Lashuel H, Iwatsubo T, Lindquist S, Aebischer P (2008) Hsp104 antagonizes alpha-synuclein aggregation and reduces dopaminergic degeneration in a rat model of Parkinson disease. J Clin Invest 118:3087–3097
337. Falsone SF, Kungl AJ, Rek A, Cappai R, Zangger K (2009) The molecular chaperone Hsp90 modulates intermediate steps of amyloid assembly of the Parkinson-related protein alpha-synuclein. J Biol Chem 284:31190–31199

338. Putcha P, Danzer KM, Kranich LR, Scott A, Silinski M, Mabbett S, Hicks CD, Veal JM, Steed PM, Hyman BT, McLean PJ (2010) Brain-permeable small-molecule inhibitors of Hsp90 prevent alpha-synuclein oligomer formation and rescue alpha-synuclein-induced toxicity. J Pharmacol Exp Ther 332:849–857
339. Daturpalli S, Wang S, Buell A, Waudby C, Meehan S, Jackson SE (2013) Hsp90 inhibits alpha-synuclein aggregation by interacting with soluble oligomers. J Mol Biol 425:4614–4628
340. Hebron ML, Lonskaya I, Moussa CE (2013) Nilotinib reverses loss of dopamine neurons and improves motor behavior via autophagic degradation of alpha-synuclein in Parkinson's disease models. Hum Mol Genet 22:3315–3328
341. Lonskaya I, Desforges NM, Hebron ML, Moussa CE (2013) Ubiquitination Increases Parkin activity to promote autophagic alpha-synuclein clearance. PLoS One 8:e83914
342. Spencer B, Potkar R, Trejo M, Rockenstein E, Patrick C, Gindi R, Adame A, Wyss-Coray T, Masliah E (2009) Beclin 1 gene transfer activates autophagy and ameliorates the neurodegenerative pathology in alpha-synuclein models of Parkinson's and Lewy body diseases. J Neurosci 29:13578–13588
343. Sarkar S, Perlstein EO, Imarisio S, Pineau S, Cordenier A, Maglathlin RL, Webster JA, Lewis TA, O'Kane CJ, Schreiber SL, Rubinsztein DC (2007) Small molecules enhance autophagy and reduce toxicity in Huntington's disease models. Nat Chem Biol 3:331–338
344. Decressac M, Mattsson B, Weikop P, Lundblad M, Jakobsson J, Bjorklund A (2013) TFEB-mediated autophagy rescues midbrain dopamine neurons from alpha-synuclein toxicity. Proc Natl Acad Sci U S A 110:E1817–E1826
345. Shimura H, Schlossmacher MG, Hattori N, Frosch MP, Trockenbacher A, Schneider R, Mizuno Y, Kosik KS, Selkoe DJ (2001) Ubiquitination of a new form of alpha-synuclein by parkin from human brain: implications for Parkinson's disease. Science 293:263–269
346. Chung KK, Zhang Y, Lim KL, Tanaka Y, Huang H, Gao J, Ross CA, Dawson VL, Dawson TM (2001) Parkin ubiquitinates the alpha-synuclein-interacting protein, synphilin-1: implications for Lewy-body formation in Parkinson disease. Nat Med 7:1144–1150
347. Xie YY, Zhou CJ, Zhou ZR, Hong J, Che MX, Fu QS, Song AX, Lin DH, Hu HY (2010) Interaction with synphilin-1 promotes inclusion formation of alpha-synuclein: mechanistic insights and pathological implication. FASEB J 24:196–205
348. Alvarez-Castelao B, Castano JG (2011) Synphilin-1 inhibits alpha-synuclein degradation by the proteasome. Cell Mol Life Sci 68:2643–2654
349. Engelender S, Kaminsky Z, Guo X, Sharp AH, Amaravi RK, Kleiderlein JJ, Margolis RL, Troncoso JC, Lanahan AA, Worley PF, Dawson VL, Dawson TM, Ross CA (1999) Synphilin-1 associates with alpha-synuclein and promotes the formation of cytosolic inclusions. Nat Genet 22:110–114
350. Casadei N, Pohler AM, Tomas-Zapico C, Torres-Peraza J, Schwedhelm I, Witz A, Zamolo I, De Heer R, Spruijt B, Noldus LP, Klucken J, Lucas JJ, Kahle PJ, Kruger R, Riess O, Nuber S (2013) Overexpression of synphilin-1 promotes clearance of soluble and misfolded alpha-synuclein without restoring the motor phenotype in aged A30P transgenic mice. Hum Mol Genet 23:767–781
351. Tanaka M, Kim YM, Lee G, Junn E, Iwatsubo T, Mouradian MM (2004) Aggresomes formed by alpha-synuclein and synphilin-1 are cytoprotective. J Biol Chem 279:4625–4631
352. Hernandez-Vargas R, Fonseca-Ornelas L, Lopez-Gonzalez I, Riesgo-Escovar J, Zurita M, Reynaud E (2011) Synphilin suppresses alpha-synuclein neurotoxicity in a Parkinson's disease Drosophila model. Genesis 49:392–402
353. Smith WW, Liu Z, Liang Y, Masuda N, Swing DA, Jenkins NA, Copeland NG, Troncoso JC, Pletnikov M, Dawson TM, Martin LJ, Moran TH, Lee MK, Borchelt DR, Ross CA (2010) Synphilin-1 attenuates neuronal degeneration in the A53T alpha-synuclein transgenic mouse model. Hum Mol Genet 19:2087–2098

354. Swinnen E, Buttner S, Outeiro TF, Galas MC, Madeo F, Winderickx J, Franssens V (2011) Aggresome formation and segregation of inclusions influence toxicity of alpha-synuclein and synphilin-1 in yeast. Biochem Soc Trans 39:1476–1481
355. Krenz A, Falkenburger BH, Gerhardt E, Drinkut A, Schulz JB (2009) Aggregate formation and toxicity by wild-type and R621C synphilin-1 in the nigrostriatal system of mice using adenoviral vectors. J Neurochem 108:139–146
356. Spencer B, Michael S, Shen J, Kosberg K, Rockenstein E, Patrick C, Adame A, Masliah E (2013) Lentivirus mediated delivery of neurosin promotes clearance of wild-type alpha-synuclein and reduces the pathology in an alpha-synuclein model of LBD. Mol Ther 21:31–41
357. Aharon-Peretz J, Rosenbaum H, Gershoni-Baruch R (2004) Mutations in the glucocerebrosidase gene and Parkinson's disease in Ashkenazi Jews. N Engl J Med 351:1972–1977
358. Sidransky E, Nalls MA, Aasly JO, Aharon-Peretz J, Annesi G, Barbosa ER, Bar-Shira A, Berg D, Bras J, Brice A, Chen CM, Clark LN, Condroyer C, De Marco EV, Durr A, Eblan MJ, Fahn S, Farrer MJ, Fung HC, Gan-Or Z, Gasser T, Gershoni-Baruch R, Giladi N, Griffith A, Gurevich T, Januario C, Kropp P, Lang AE, Lee-Chen GJ, Lesage S, Marder K, Mata IF, Mirelman A, Mitsui J, Mizuta I, Nicoletti G, Oliveira C, Ottman R, Orr-Urtreger A, Pereira LV, Quattrone A, Rogaeva E, Rolfs A, Rosenbaum H, Rozenberg R, Samii A, Samaddar T, Schulte C, Sharma M, Singleton A, Spitz M, Tan EK, Tayebi N, Toda T, Troiano AR, Tsuji S, Wittstock M, Wolfsberg TG, Wu YR, Zabetian CP, Zhao Y, Ziegler SG (2009) Multicenter analysis of glucocerebrosidase mutations in Parkinson's disease. N Engl J Med 361:1651–1661
359. Mazzulli JR, Xu YH, Sun Y, Knight AL, McLean PJ, Caldwell GA, Sidransky E, Grabowski GA, Krainc D (2011) Gaucher disease glucocerebrosidase and alpha-synuclein form a bidirectional pathogenic loop in synucleinopathies. Cell 146:37–52
360. Sardi SP, Clarke J, Kinnecom C, Tamsett TJ, Li L, Stanek LM, Passini MA, Grabowski GA, Schlossmacher MG, Sidman RL, Cheng SH, Shihabuddin LS (2011) CNS expression of glucocerebrosidase corrects alpha-synuclein pathology and memory in a mouse model of Gaucher-related synucleinopathy. Proc Natl Acad Sci U S A 108:12101–12106
361. Brockmann K, Srulijes K, Hauser AK, Schulte C, Csoti I, Gasser T, Berg D (2011) GBA-associated PD presents with nonmotor characteristics. Neurology 77:276–280
362. Sardi SP, Clarke J, Viel C, Chan M, Tamsett TJ, Treleaven CM, Bu J, Sweet L, Passini MA, Dodge JC, Yu WH, Sidman RL, Cheng SH, Shihabuddin LS (2013) Augmenting CNS glucocerebrosidase activity as a therapeutic strategy for Parkinsonism and other Gaucher-related synucleinopathies. Proc Natl Acad Sci U S A 110:3537–3542
363. Yap TL, Gruschus JM, Velayati A, Westbroek W, Goldin E, Moaven N, Sidransky E, Lee JC (2011) Alpha-synuclein interacts with glucocerebrosidase providing a molecular link between Parkinson and Gaucher diseases. J Biol Chem 286:28080–28088
364. Sevlever D, Jiang P, Yen SH (2008) Cathepsin D is the main lysosomal enzyme involved in the degradation of alpha-synuclein and generation of its carboxy-terminally truncated species. Biochemistry 47:9678–9687
365. Chu Y, Dodiya H, Aebischer P, Olanow CW, Kordower JH (2009) Alterations in lysosomal and proteasomal markers in Parkinson's disease: relationship to alpha-synuclein inclusions. Neurobiol Dis 35:385–398
366. Cullen V, Lindfors M, Ng J, Paetau A, Swinton E, Kolodziej P, Boston H, Saftig P, Woulfe J, Feany MB, Myllykangas L, Schlossmacher MG, Tyynela J (2009) Cathepsin D expression level affects alpha-synuclein processing, aggregation, and toxicity in vivo. Mol Brain 2:5
367. Krumova P, Meulmeester E, Garrido M, Tirard M, Hsiao HH, Bossis G, Urlaub H, Zweckstetter M, Kugler S, Melchior F, Bahr M, Weishaupt JH (2011) Sumoylation inhibits alpha-synuclein aggregation and toxicity. J Cell Biol 194:49–60
368. Kim YM, Jang WH, Quezado MM, Oh Y, Chung KC, Junn E, Mouradian MM (2011) Proteasome inhibition induces alpha-synuclein SUMOylation and aggregate formation. J Neurol Sci 307:157–161

Top Med Chem (2016) 18: 111–148
DOI: 10.1007/7355_2014_69
© Springer-Verlag Berlin Heidelberg 2014
Published online: 9 December 2014

Leucine-Rich Repeat Kinase 2 (LRRK2) Inhibitors

Paul Galatsis, Jaclyn L. Henderson, Bethany L. Kormos, and Warren D. Hirst

Abstract Mutations in the leucine-rich repeat kinase 2 (LRRK2) are the most common known cause of autosomal dominant Parkinson's disease (PD), accounting for approximately 1% of "sporadic" and 4% of familial cases. These mutations either lead directly to an increased kinase activity (G2019S and I2020T are in the kinase activation loop) or to a reduced GTPase activity (R1441C/G and Y1699C), that in turn positively regulate kinase activity. The physiological substrate of the LRRK2 kinase has yet to be definitively identified, yet autophosphorylation is emerging as a relatively robust measure of its activity. LRRK2 has been implicated in a number of diverse cellular processes such as vesicular trafficking, microtubule dynamics, protein translation control, inflammation, and immune function, all of which have been linked to PD. LRRK2 is a large, multi-domain protein; a thorough understanding of the protein domain organization and identification of interacting partners is important to determine the underlying mechanism of LRRK2. Substantial recent effort has been directed towards identifying potent LRRK2 kinase inhibitors, from the repurposed kinase inhibitors to the first through third generation of LRRK2-focused kinase inhibitors, from a range of chemotypes, which are now providing researchers with new tools to better interrogate LRRK2 function.

Keywords Inhibitors, Kinase, Leucine-rich repeat kinase 2, LRRK2

P. Galatsis (✉) and J.L. Henderson
Worldwide Medicinal Chemistry, Neuroscience Chemistry, Pfizer Worldwide Research and Development, 610 Main St., Cambridge, MA 02139, USA
e-mail: paul.galatsis@pfizer.com

B.L. Kormos
Worldwide Medicinal Chemistry, Computational Chemistry, Pfizer Worldwide Research and Development, 610 Main St., Cambridge, MA 02139, USA

W.D. Hirst
Neuroscience Research Unit, Pfizer Worldwide Research and Development, 610 Main St., Cambridge, MA 02139, USA

Contents

Abbreviations

6-OHDA	6-Hydroxydopamine
ANK/AR	Ankyrin repeats
ARM	Armadillo
BA	Brain availability
BBB	Blood brain barrier
BI	Brain impairment
CNS	Central nervous system
COR	C-terminal of Roc
GPCR	G-protein coupled receptor
GWAS	Genome-wide association study
HB	hydrogen bond
HBD	hydrogen bond donor count
hERG	Human ether-a-go-go-related gene
HLM	Human liver microsome
KO	Knockout
LE	Ligand efficiency
LipE	Lipophilic ligand efficiency
logD	Natural logarithm of the distribution coefficient
logP	Natural logarithm of the water/octanol partition coefficient
LPS	Lipopolysaccharide
LRR	Leucine-rich repeats
LRRK2	Leucine-rich repeat kinase 2
MDR1	Multidrug resistance
MPO	Multiparameter optimization
MPTP	1-Methyl-4-phenyl-1,2,3,6-tetrahydropyridine
MW	Molecular weight

NHP	Nonhuman primate
PBMC	Peripheral blood mononuclear cell
PD	Parkinson's disease
PSA	Polar surface area
RLM	Rat liver microsome
Roc	Ras of complex proteins
ROS	Reactive oxygen species
RRCK	Ralph Russ Canine Kidney
SAR	Structure–activity relationship
SN	Substantia nigra
THLE	Transformed human liver epithelial
VDW	Van der Waals
WD40	WD40 repeats
α-syn	α-Synuclein

1 Introduction

Parkinson's disease (PD) is the most common movement disorder and the second most common neurodegenerative disorder. The etiology of PD is complex but the most common phenotype is the loss of dopaminergic neurons of the substantia nigra (SN) leading to the clinical symptoms of bradykinesia, resting tremors, rigidity, and postural instability. Historically, at a pathological level, PD is characterized by inclusions containing the synaptic protein α-synuclein (α-syn) in the cell bodies and processes of surviving neurons (known as Lewy bodies and Lewy neurites, respectively). However, more recent observations describe not only a wider distribution of α-syn pathology, but also accumulation of tau and Aβ aggregates, suggesting a more complex pathology [1]. In addition, inflammation, including reactive gliosis, is observed in the striatum and substantia nigra of PD patients [2, 3].

The penetrance of the disease increases with age with 1% of the population over 65 being affected, rising to 5% by age 85 [4]. The past 17 years has led to considerable progress in both the identification of mutations that cause disease and in the mapping of common variants that alter risk for PD [5]. It is now clearly established that many, if not all, forms of Parkinson's disease (PD) contain a genetic component [5], and mutations in the leucine-rich repeat kinase 2 (LRRK2) are the most common known cause of autosomal dominant PD, accounting for approximately 1% of "sporadic" and 4% of familial cases [5–7]. In specific populations, notably Ashkenazi Jews and North African Berber Arabs, the prevalence of LRRK2 mutations can be as high as 40% [8]. Of the LRRK2 mutations, the most common is G2019S with penetrance ranging from ~30 to 70% by the age of 80 [8–10]. This mutation occurs in the kinase domain of the protein, increasing the kinase activity, and consequently, a great deal of effort has been directed towards identifying potent LRRK2 kinase inhibitors, with a biopharmaceutical profile congruent with clinical evaluation, as novel therapeutics for PD. This chapter describes the current state of the art in developing appropriate kinase inhibitors.

2 LRRK2 Biology and Pharmacology

2.1 LRRK2 Genetics and Human Biology

The LRRK2 gene has 51 exons, with multiple potential splice sites, which encodes a large, 2,527 amino acid protein containing two predicted enzymatic domains (GTPase and kinase) and multiple protein–protein interaction domains (Fig. 1). See below for a more detailed account of LRRK2 structural biology (Sect. 3). A number of novel variants have been identified in this gene in PD patients, but only seven of these (N1437H, R1441C, R1441G, S1761R, Y1699C, G2019S, and I2020T) can be considered as definitively disease causing, on the basis of co-segregation with disease in families, and an absence in controls [5, 11–13]. These mutations either lead to an increased kinase activity (G2019S and I2020T) or to a reduced GTPase activity (R1441C/G and Y1699C), which in turn regulates kinase activity [14–16] (*vide infra*).

The G2019S mutation is relatively frequent in some populations from Southern Europe and in certain populations, such as the Ashkenazi Jews and North African Berber Arabs, where the prevalence can be as high as 40% [8]. However, there is an incomplete but age-related penetrance, ranging from ~30 to 70% by 80 years old in different studies, that has been estimated for carriers of the G2019S mutation, and the associated range of PD onset age is broad, including patients with early and late disease onset [8–10]. Dopaminergic neuronal loss and gliosis in the substantia nigra are the common pathological features in patients with LRRK2 mutations, and classical Lewy bodies are found in the majority of them. However, in some cases α-synuclein-positive inclusions are not observed, and only tau-positive or ubiquitin-positive inclusions are seen [7]. Overall, the clinical characteristics of patients with LRRK2 mutations, particularly those with the more common G2019S mutation, are very similar to those with sporadic PD [8, 17].

The incomplete penetrance clearly indicates that other factors are involved in the pathogenesis (Fig. 1); these may include other genetic contributions, both known (e.g., α-synuclein, Tau, RAB7L1, GAK) [5, 18] and yet to be defined, and/or "environmental" factors such as inflammation or oxidative/nitrative/unfolded and/or misfolded protein stress [19–21], many of which are actively being investigated. For example, LRRK2 kinase levels are increased by inflammatory mediators, such as IFNγ and LPS, in vitro [22, 23] and in vivo [23], and two recent reports suggest that LRRK2 levels are increased in sporadic PD brains [24, 25], potentially regulated by microRNA-205 [24]. LRRK2 has been shown to exist as a dimeric protein [26–29] with this form having greater kinase activity [28, 30]. These data have led to a proposed model in which LRRK2 cycles from a cytoplasmic, low-activity monomer to a higher-activity plasma membrane-associated dimer driven by, for example, LPS [28, 31, 32]. However, such biochemical experiments, investigating these potentially more active conformations of LRRK2, have not yet been performed in postmortem patient brains. Furthermore, to date, there is no direct evidence of increased LRRK2 kinase activity in postmortem patient brains from either G2019S mutation carriers or patients with sporadic disease. This is partly due to the challenges of immunoprecipitating sufficient active LRRK2 for in vitro phosphorylation assays, compounded by

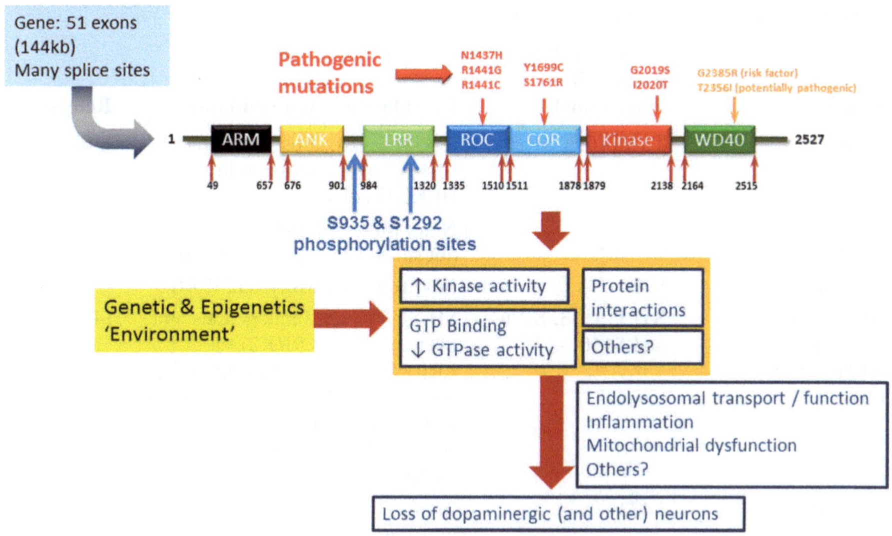

Fig. 1 LRRK2 protein domain structure and putative PD pathways

the potential effects of postmortem delay on the activity [33]. The lack of a validated substrate (vide infra) and the reliance on surrogate substrate, e.g., LRRKtide and NICtide [34, 35], also hampered such efforts. Despite this we, and others, are currently investigating these assays, and other end-points, to provide a critical link, which is currently missing, between LRRK2 and sporadic disease.

2.2 *LRRK2 Substrates*

The physiological substrate of the LRRK2 kinase has yet to be definitively identified. A number of potential substrates, other than LRRK2 (see next section), have been proposed (Table 1). However, full validation of these, including a clear demonstration of an increase in human PD patient cells and tissues and reduction in phosphorylation with multiple, structurally diverse inhibitors and by knocking out LRRK2, remains to be demonstrated.

2.3 *Effects of Mutations on Kinase and GTPase Activity*

The G2019S mutation, in the activation loop, has been consistently shown to result in increased kinase activity of LRRK2 [57, 58]. However, the functional consequences of the other mutations reported in the literature are conflicting. The other kinase domain mutation (I2020T) has been reported to either increase [59] or decrease kinase activity [34, 60]. Similarly, mutations in the GTPase domain have been demonstrated

Table 1 Putative LRRK2 kinase substrates

Protein	Phosphorylation site(s) – if determined	Identification and validation	References
4E-BP	T37/T46, T70	In vitro assay, drosophila, and LRRK2 overexpression in HEK293 cells	[36]
		Note: follow-up publications did not show robust phosphorylation	[37, 38]
α-Synuclein	S129	In vitro kinase assay (G2019S)	[39]
Akt1 (PKB), ARHGEF7, ARHGEF11 and others	S473, S153, S176 and S150	SH-SY5Y cells (endogenous and overexpressed LRRK2) +/– LRRK2-IN-1. Pathway analysis, microglial inflammatory responses and neurite outgrowth demonstrated off-target effects of LRRK2-IN-1	[40]
Akt1 (PKB)	S473	In vitro kinase assay and LRRK2 knockdown	[41]
ArfGAP1	S155, S246, S284, T189, T216, T292	Interaction studies and in vitro kinase assay	[42, 43]
β-tubulin	T107	In vitro assay and LRRK2 overexpression in HEK293 cells	[44]
Endophilin A	S75	In vitro assay and effects on synaptic vesicle trafficking	[45]
Ezrin/radixin/moesin proteins	Not determined	Primary neurons pERM increased with G2019S	[46]
FoxO	S259 (drosophila, corresponding human residue: S319)	In vitro assay and drosophila	[47]
MAP2	T1433	Consensus phosphorylation motif and peptide phosphorylation	[48]
MARKK	Not determined	In vitro kinase assay (G2019S)	[49]
MKK3/4/6/7	K149/S207 (MKK4 and 7)	In vitro kinase assay (G2019S)	[50, 51]
MKK3/6/7	Not determined	In vitro kinase assay	[52]
Moesin	T558	KESTREL screen and in vitro kinase assay	[34]
Praja	T74	Consensus phosphorylation motif and peptide phosphorylation	[48]
Ribosomal protein s15	T136	Interacting phosphoproteins identified by tandem affinity purification, LRRK2 overexpression in neurons	[53]
RIPK2	Not determined	Protein array with G2019S	[49]

(continued)

Table 1 (continued)

Protein	Phosphorylation site(s) – if determined	Identification and validation	References
Serine–protein kinase ATM (ataxia telangiectasia mutated)	T1769	Consensus phosphorylation motif and peptide phosphorylation	[48]
Snapin	T117	Interaction identified by yeast-2-hybrid. In vitro kinase assays with purified recombinant Snapin and LRRK2	[54]
Tau (tubulin associated)	T181	In vitro kinase assay and LRRK2 knockdown and overexpression in SH-SY5Y cells	[55]
	T149, T153, T205, and S199/S202/T205	In vitro tau phosphorylation; LRRK2/Tau_{P301L} transgenic mice	[56]
ULK1	T456	Consensus phosphorylation motif and peptide phosphorylation	[48]

to increase kinase activity [57, 61], whereas in other studies, they have had little or no effect [34, 58]. More recent data have shown that the R1441C/G and Y1699C mutations reduced the GTPase activity that in turn regulates kinase activity [14, 15]. The majority of studies have used either recombinantly expressed and purified LRRK2 [58] or immunoprecipitated LRRK2 from recombinant mammalian expression systems [34, 57, 61] and investigated autophosphorylation or the use of surrogate substrates (myelin basic protein, LRRKtide, NICtide), all which are relatively poor substrates under those assay conditions.

LRRK2's autophosphorylation has been characterized [30, 62, 63], and given the caveats of the validation of the exogenous substrate, perhaps this end-point is currently the most relevant. This is supported by recent data that demonstrated an increase in phosphorylation at S1292 by the pathological mutations [16]. Importantly, and in contrast to most of the previous results using purified enzyme, the magnitude of increase in phosphorylation of S1292 by, for example, G2019S is over 10-fold, when compared to wild-type LRRK2 [16] versus the 2–3-fold typically observed with the isolated enzyme [58]. This is clearly suggesting that the microenvironment of LRRK2 within the cell, including accessory proteins, interaction partners, and membrane association, is critical for the optimal kinase activity and that such a measure of kinase activity is a potentially valuable biomarker if it can be applied to patient samples. This will be a challenge as the stoichiometry of phosphorylation at S1292 is very low ($<1\%$) ([16], unpublished observations), which may also reflect the tight control of LRRK2's activity in the cellular context.

There are additional phosphorylation sites on LRRK2, including S910 and S935 [64], which have been shown to interact with 14-3-3 proteins and regulate the subcellular localization of LRRK2 [65]. While not autophosphorylation sites, they are clearly sensitive to LRRK2 kinase inhibitors, potentially due to conformational changes [66] exposing these residues to phosphatases, and have been used extensively in both cellular and in vivo studies (vide infra).

2.4 Role of LRRK2 in Normal and Pathological Biological Pathways

The understanding of LRRK2's role in normal and pathological biological pathways is still at a relatively early stage, with a number of key questions that remain to be answered. For example, the endogenous substrate for the kinase domain is not known (vide supra and Table 1), and despite recent progress identifying interaction partners [67, 68], the full understanding of LRRK2's function and interactome in different cell types remains an area of active research. The signaling pathways, through which LRRK2 elicits its actions and others that it potentially modulates, are emerging [69, 70] and its role in PD pathophysiology, while not definitive, is also developing.

LRRK2 has been implicated in a number of diverse cellular processes [19, 71, 72] including autophagy, vesicular trafficking, microtubule dynamics, neurite outgrowth, endosomal/synaptic dysregulation, protein translation control, mitochondrial pathology/ROS, WNT and MAPK/MEK/ERK/EIF2 signaling pathways, inflammation, and immune function, all of which have been linked to PD.

2.5 Animal Models to Understand LRRK2 Function

Studies in *Drosophila melanogaster* [73] and *Caenorhabditis elegans* [74] have provided important insights into LRRK2 toxicity in these species. The LRRK2 transgenic mouse models created to date do not completely recapitulate the hallmarks of PD (i.e., dopaminergic neuronal loss, α-synuclein accumulation, the development of Lewy bodies, and behavioral phenotype) [75–77], and with the exception of pharmacodynamic end-points, measuring reduction of LRRK2 phosphorylation at S935 and S1292 has not been used, to date, for any long-term LRRK2 inhibitor studies [16]. In contrast, viral overexpression of LRRK2 does result in dopaminergic neuronal loss [78, 79], but these models need to be fully validated with the more selective, brain-penetrant compounds described herein.

Toxins such as 6-hydroxydopamine (6-OHDA) or 1-methyl-4-phenyl-1,2,3,6-tetrahydropyridine (MPTP) have been effective in acutely inducing DA cell loss [80]. However, the LRRK2 knockout mouse displays the same sensitivity to MPTP as wild-type mice [81] suggesting that LRRK2 may not be involved in the events downstream of the mitochondrial toxicity in the mouse brain. There are no published reports on 6-OHDA in in vivo models, with LRRK2 overexpression or with endogenous LRRK2. The use of other toxins, such as LPS, may provide a more relevant model as LRRK2 kinase levels are increased by LPS in vitro [22, 23] and in vivo [23].

The lack of validated mammalian preclinical models of LRRK2 has been an impediment to the development of LRRK2 kinase inhibitors. This has also been hampered by the lack of suitable tool compounds that possess the appropriate pharmacokinetic properties and safety profiles for long-term dosing studies that are likely to be required.

2.6 LRRK2 and Potential Safety Concerns

LRRK2 is widely expressed, with highest expression in kidney, lung, and peripheral blood monocytes and lower levels in the brain [82]. LRRK2 functions in the peripheral tissues are also not well understood, and transgenic animals, knock-ins and knockouts, have been generated in an attempt to gain greater insight into LRRK2 biology. In all cases, the animals are viable and exhibit normal life spans. Initial studies with LRRK2 knockout (KO) mice exhibited phenotypes which showed a lack of hypersensitivity to MPTP, caused impairment of protein degradation pathways, resulted in accumulation of α-synuclein in kidney, not brain, and resulted in apoptotic cell death [76, 80, 83]. A single report with conditional LRRK2 G2019S overexpression in rats showed impaired dopamine reuptake and improved locomotor activity without loss of SN dopaminergic neurons [84]. LRRK2 KO rats are now commercially available [85]. While these animals appear to not have functional impairment, they do exhibit two distinct peripheral phenotypes, similar to the LRRK2 KO mice: (1) the kidneys present with a dark color that show inclusion bodies upon microscopic examination, and (2) type II pneumocytes of the lungs show changes in lamellar body morphology with altered surfactant secretion, measured in vitro [86–89]. The changes in type 2 pneumocyte vacuolation/lamellar body enlargement have recently been observed with 7- and 29-day NHP toxicology studies [90]. It remains to be determined whether KO phenotypes are a consequence of a total lack of LRRK2 from the embryonic stage to adult and if the drug-mediated phenotypes are chemotype or mechanism based.

3 LRRK2 Structural Biology

3.1 LRRK2 Protein Domain Structure

The LRRK2 gene encodes a 286 kDa protein comprising 2,527 amino acids [6, 91]. LRRK2 protein belongs to the Roco family of proteins whose members contain multiple domains including a tandem Roc (Ras of complex proteins)–COR (C-terminal of Roc) motif [92, 93]. As with the other members of the Roco family, LRRK2 contains multiple domains (Fig. 1): the enzymatic Roc–COR and kinase domains, which are flanked by putative protein–protein interaction domains, including the N-terminal armadillo (ARM) repeats, ankyrin repeats (AR or ANK) and leucine-rich repeats (LRR), and the C-terminal WD40 repeats [94, 95]. Currently, there is no unanimous agreement on the identity and number of armadillo, ankyrin, leucine-rich, and WD40 repeats in the LRRK2 protein [94]. A thorough understanding of the protein domain organization and identification of interacting partners is important to understand the function(s) of LRRK2.

3.1.1 Protein–Protein Interaction Domains

No X-ray crystal structures have yet been reported for any of the LRRK2 putative protein–protein interaction domains; therefore, its domain identification relies on bioinformatic analysis of the primary sequence to predict domain composition and boundaries.

The N-terminal LRRK2-specific repeats characterized by Marín [94] have been predicted to adopt the folds of 13 armadillo-type repeats from residues E49 to K657 [95]. Each ARM repeat is made up of ~42 amino acids that form three α-helices; tandem ARM repeats form a super helical structure that is involved in protein–protein interactions in other ARM repeat proteins [96]. This domain is predicted to be the largest LRRK2 domain containing ~600 amino acids.

Seven ANKs have been predicted to comprise the subsequent LRRK2 domain, made up of residues F676–S901 [95, 97]. Each AR is made up of 33 amino acids that form two antiparallel α-helices followed by a β-hairpin or a long loop, assembling to form a curved structure [98]. AR proteins are involved in a wide range of functions and typically mediate protein–protein interactions [99].

The next domain in sequence is the LRR domain, from where LRRK2 gets its name. As many as 14 LRRs have been predicted from residues I984 to R1320 [95, 100], though fewer have also been suggested [92, 94, 97, 101]. Each LRR is typically made up of 20–30 amino acids (~24 in LRRK2) and contains an unusually high number of leucine residues that form a β-strand followed by an α-helix or extended chain [102]. Tandem LRRs assemble into a horseshoe- or solenoid-type shape whose curvature depends on the specifics of the secondary structure. The framework of LRR domains makes them well suited for protein–protein interactions, and the concave surface of the LRR structure is usually, though not always, the site of macromolecular recognition or dimerization [103].

The C-terminal domain has been predicted to contain one [94], two [92], or seven [95, 97] WD40 repeats; the seven repeats are predicted to occur between residues N2164 and K2515 [95]. Each WD40 repeat contains a four-stranded, antiparallel β-sheet made up of 40 residues [97]. They usually occur in multiples of six or seven and assemble to form a propeller-like structure that serves as a rigid scaffold for protein–protein interactions [104].

3.1.2 Enzymatic Domains

The Roc domain of LRRK2 is a Ras-like GTPase domain that is followed immediately by the COR domain as in all members of the ROCO protein family

[93]. These domains approximately span residues M1335–N1510 and F1511–E1878, respectively [97]. Ras family GTPases act as molecular switches between an active GTP-bound state and an inactive GDP-bound state [105, 106]. Evolutionary analysis indicates that not all ROCO proteins contain a kinase domain; as such, it has been suggested that GTPase activity is the primary function of ROCO proteins and those with kinase domains evolved independently to regulate that function [92, 93, 95, 107].

Two X-ray crystal structures have been solved providing the ability to form hypotheses about the LRRK2 Roc and COR domains and their functions based on structural information. One is a structure of the human LRRK2 Roc domain bound to GDP-Mg^{2+}, which reveals a domain-swapped homodimer [108]. This structure suggests the Roc dimer acts as one functional unit with each nucleotide binding site formed by contributions from both monomers. The other relevant X-ray crystal structure contains both the Roc and COR domains from the bacteria *Chlorobium tepidum* and does not exhibit the domain swapping configuration [109]. This structure suggests that the COR domain is responsible for dimerization, that the Roc domain is stabilized by the COR domain, and that the Roc–COR domain is stabilized by nucleotide binding. Much discussion of the validity and relevance of these X-ray crystal structures has taken place [105, 107, 110–114]. Regardless, these structures both have implications for LRRK2 GTPase function that have spurred important discussions, and results from hypotheses generated based on these structures will only improve our understanding of the LRRK2 protein.

The kinase domain has the most straightforward therapeutic potential and thus has garnered the most interest of all of the domains in the LRRK2 protein. Characterized as a Ser/Thr kinase in the TKL group of kinases, it has the highest homology to LRRK1 and the receptor-interacting protein (RIP) kinases [94, 115], which are sensors of intracellular and extracellular stresses [116]. The LRRK2 kinase domain spans residues Q1879–V2138. Protein kinases catalyze the transfer of a phosphoryl group from ATP to serine, threonine, or tyrosine residues in protein substrates and mediate most cellular signaling processes [117]. The physiological substrate of the LRRK2 kinase has yet to be definitively identified (vide supra). The X-ray crystal structure of the kinase domain from the slime mold *Dictyostelium discoideum* Roco4 protein may provide insight into the structure of the LRRK2 kinase domain [118]. With a kinase domain that is ~30% identical and ~50% similar to that of LRRK2, as well as a similar domain structure that contains LRRs, Roc–COR, kinase, and WD40 repeat domains, the *D. discoideum* Roco4 kinase domain may be a suitable crystallography surrogate to help build an understanding of LRRK2 kinase structure and mechanism.

3.2 LRRK2 Pathogenic PD Mutations

To date, the proven pathogenic PD mutations (Fig. 1) segregate to the enzymatic Roc–COR and kinase domains of LRRK2 [13, 119, 120], though there are a number of variants that have been identified as PD risk factors that are distributed

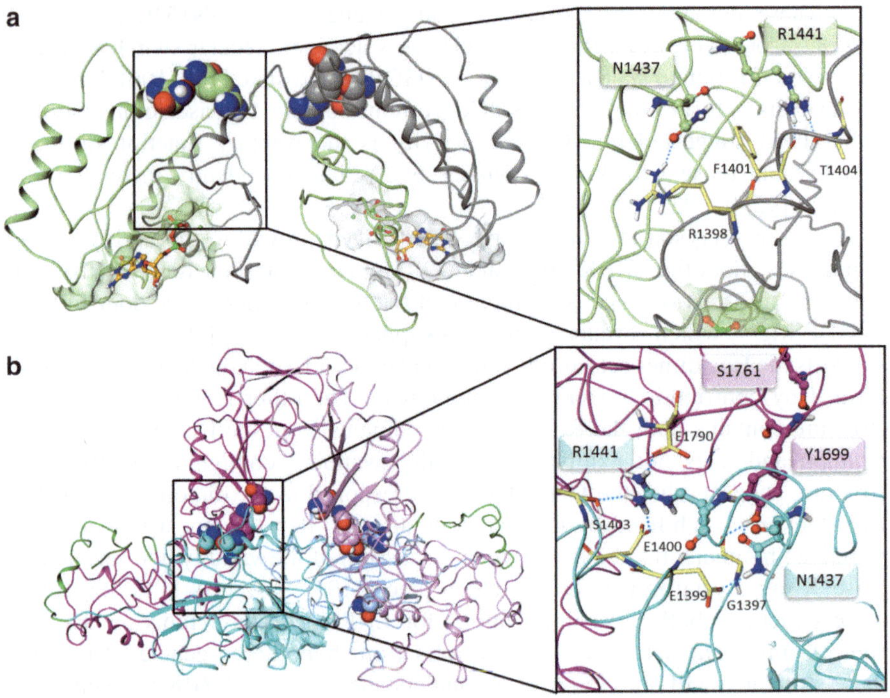

Fig. 2 Roc–COR structures with positions of pathogenic mutations and interacting residues. (**a**) Self-homology model of the domain-swapped LRRK2 Roc–Roc homodimer (PDB ID: 2ZEJ). *Orange*: GDP; *Green*: monomer A; *Gray*: monomer B; *Yellow*: interacting residues. (**b**) Homology model of LRRK2 Roc–COR dimer based on the *C. tepidum* structure (PDB ID: 3DPU). *Cyan*: Roc monomer A; *Pink*: COR monomer A; *Light blue*: Roc monomer B; *Light pink*: COR monomer B; *Yellow*: interacting residues

throughout the entire protein [121, 122]. As such, these mutations have the potential to disrupt not only LRRK2 enzymatic functions but also protein–protein interactions. In the absence of LRRK2 X-ray crystal structures, homology models [123, 124] have been valuable tools to help understand the position of pathogenic mutations and the effect(s) they may have on the function of the LRRK2 protein.

3.2.1 Roc–COR Pathogenic Mutations

The R1441C/G/H and N1437H pathogenic mutations are located in the Roc domain and the Y1699C and S1761R mutations are located in the COR domain. The two current models for the Roc domain suggest different roles for these residues (Fig. 2) [125]. In the domain-swapped homodimer X-ray crystal structure of the human LRRK2 Roc domain proposed by Cookson and coworkers [108], R1441 and N1437 are situated such that they stabilize the Roc–Roc dimer interface through interactions with the other monomer. R1441 forms HB interactions with the backbone carbonyl of

F1401 and the side chain of T1404 and forms a cation-π interaction with the side chain of F1401, while N1437 forms a HB interaction with the side chain of R1398.

Alternatively, in the X-ray crystal structure of the Roc–COR dimer from *C. tepidum* proposed by Wittinghofer and coworkers [109], the residues equivalent to R1441, N1437, and Y1699 (Y558$^{C.\ tepidum}$, H554$^{C.\ tepidum}$, and Y804$^{C.\ tepidum}$, respectively) are clustered together at the interface of the Roc and COR domains, involved in interactions within and between the Roc and COR domains. Based on the Y558$^{C.\ tepidum}$ interactions, it is hypothesized that the aliphatic side chain of R1441 is involved in structurally important hydrophobic interactions with the COR domain. A homology model based on this structure suggests that R1441 may also form HB interactions with the side chains of E1400 and S1403 in the Roc domain and that N1437 forms HB interaction with the side chains of E1399 in the Roc domain and Y1699 in the COR domain [126] (Fig. 2a). Based on the Y804$^{C.\ tepidum}$ interactions, Y1699 may form a hydrogen bond with N1437 (H554$^{C.\ tepidum}$). A homology model based on this structure suggests Y1699 may also form a HB with the N1437 side chain or the G1397 backbone carbonyl and a CH-π interaction with the C^{α} of N1437 [126] (Fig. 2b).

Neither model places these residues close enough to the binding site to interact with GTP or GDP, so the pathogenic mutations do not appear to disrupt nucleotide binding directly. In both models, mutation of R1441 to Cys, Gly, or His and N1437 mutation to His would disrupt HB and van der Waals (VDW) interactions, and in the *C. tepidum* model, mutation of Y1699 to Cys would disrupt a HB between the Roc and COR domains. Therefore, evidence from both models suggests that these residues play an important role in dimer formation or interactions between domains, affecting LRRK2 protein function (vide supra). In addition, these three residues form interactions with residues in the Switch II region of the Roc domain [109], a conserved region in Ras proteins that is important for GTPase function [127, 128]. This suggests that mutation of these residues could also affect LRRK2 GTPase function by altering interactions with the Switch II region.

The S1761 residue in the COR domain corresponds to S852$^{C.\ tepidum}$ in the *C. tepidum* structure. This residue is farther away from the Roc–COR interface and Switch II than the other pathogenic mutations and is situated on the loop between β9 and β10. As such, it may be involved in protein–protein interactions with other LRRK2 domains or other scaffolding proteins. Mutation to Arg could affect these protein–protein interactions or the flexibility of the loop.

3.2.2 Kinase Domain Pathogenic Mutations

The G2019S and I2020T pathogenic mutations are located in the kinase domain. In the absence of an X-ray crystal structure of the LRRK2 kinase domain, a number of homology models have been reported and used to understand the position of these residues and the implications of the pathogenic mutations. There is no consensus on a "best" template to build LRRK2 kinase domain homology models due to the fact that the kinases for which X-ray crystal structures have been published have at most ~30% sequence identity to LRRK2. The most common templates have been from

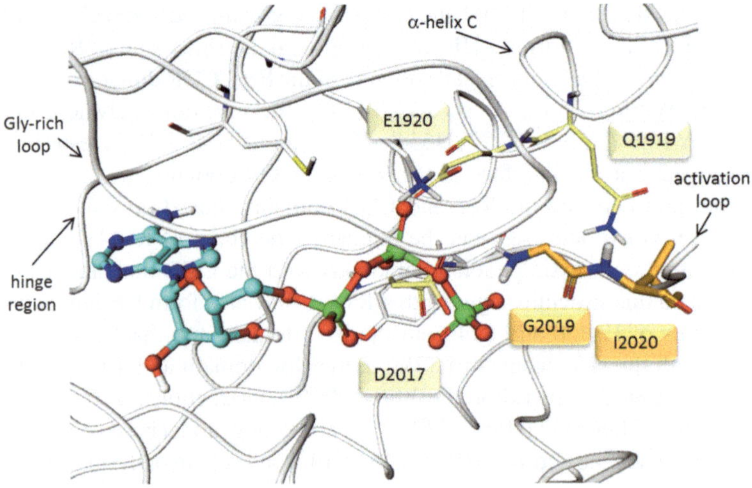

Fig. 3 LRRK2 ATP-binding site. ATP (*cyan*) is shown docked into a LRRK2 homology model based on the X-ray crystal structure of JAK3 with CP-690550 (PDB ID: 3LXK). *Orange*: location of the G2019S and I2020T mutations; *Yellow*: hypothesized interacting residues

B-Raf [94, 101, 129–131], JAK-2 [132–138], MLK1 [132, 136–139], ROCO4 [131, 140], and TAK1 [132, 141, 142] kinases, chosen based on overall kinase sequence identity, ATP-binding site identity, and/or crossover of inhibitor activity. Consensus from these alignments is that G2019 is the Gly in the conserved DFG motif (DYG in LRRK2) and I2020 is the subsequent residue in the activation loop (Fig. 3) [125].

Modeling of the G2019S mutation suggests that the change from a flexible Gly residue to a Ser causes the activation loop to be less flexible due to an increase in HB interactions with D2017, E1920, or other nearby residues, increasing the population of conformations in the active state [114, 131, 136, 141]. The X-ray crystal structure of Roco4 with the mutation that corresponds to LRRK2 G2019S, G1179S^{Roco4}, was solved to 2.04 Å resolution and reveals a HB interaction between G1179S^{Roco4} and R1077^{Roco4} in the αC-helix, which corresponds to Q1919 in LRRK2 [118]. This construct was shown to have increased kinase activity compared to WT. In addition, the double mutants G1179S^{Roco4}/R1077A^{Roco4} in Roco4 and G2019S/Q1919A in LRRK2, in which the Ser is unable to make a HB interaction to the αC-helix to stabilize the active conformation, have nearly wild-type activity. Metadynamic simulations combined with kinetics studies suggested that the energy barrier to achieve the inactive DYG-out conformation in LRRK2 is much higher for the G2019S mutant than for WT [131]. Together, this evidence suggests that an additional hydrogen bond in the G2019S mutant stabilizes the active kinase conformation, causing an increase in kinase activity.

Modeling of the I2020T mutation, on the other hand, does not suggest a clear hypothesis for its pathogenicity. The X-ray crystal structure of Roco4 kinase with the corresponding mutation, L1180T ^{Roco4}, was solved to 2.3 Å resolution and shows the side chain of the Thr interacting with solvent [118]. It was suggested that

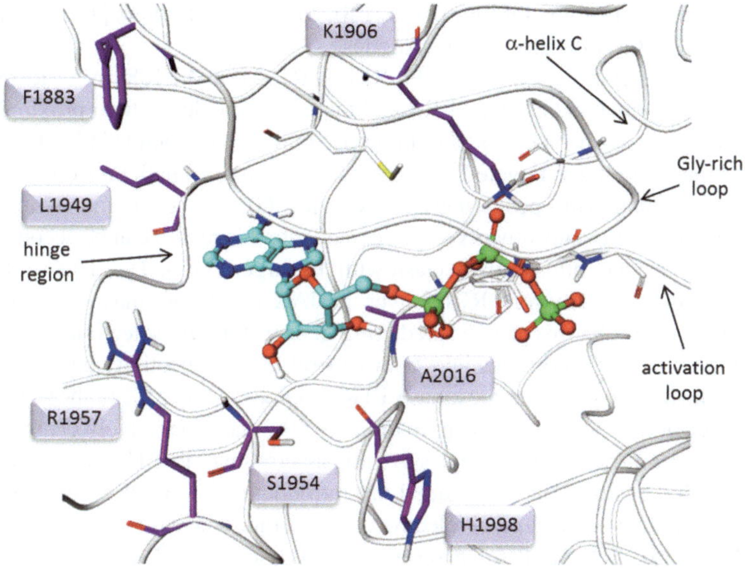

Fig. 4 LRRK2 ATP-binding site. ATP (*cyan*) is shown docked into a LRRK2 homology model based on the X-ray crystal structure of JAK3 with CP-690550 (PDB ID: 3LXK). *Purple*: residues that have been targeted for LRRK2 specificity

the effect of the mutation may only be evident upon LRRK2 dimerization or interactions with other domains, which cannot be discerned from the X-ray crystal structure of the kinase domain alone. Alternatively, molecular dynamics simulations suggest the Thr mutation may form a HB interaction with the backbone carbonyl of D2017, which is involved in substrate binding [60, 114]. Metadynamic simulations combined with kinetic studies suggested that the I2020T mutant stabilized the active DYG-in conformation in LRRK2 compared to WT [60].

3.3 *Using LRRK2 Structure to Guide Kinase Inhibitor Design*

Most of the LRRK2 kinase inhibitors that have been published to date are ATP-competitive inhibitors that contain hinge-binding motifs common to many kinase inhibitors (vide infra). Through the use of structure–activity relationships (SAR), X-ray crystal structures, and homology models, specific residues in the LRRK2 ATP-binding pocket have been targeted for inhibitor interactions to improve potency and selectivity for LRRK2 kinase (Fig. 4). Hydrophobic interactions with A2016 have been shown to be important for potency and selectivity of some LRRK2 kinase inhibitors [35, 140, 143, 144]. Selectivity of these compounds is especially improved over kinases that have a more polar residue at that position – 28% of kinases have a Ser or Thr at that position. Mutation of A2016 to Thr abolishes activity in some inhibitors [35, 140, 143], but not others [145–147], which is indicative of the inhibitor binding modes.

The non-conserved residues L1949, S1954, R1957, and F1883 were identified as potential selectivity handles in the LRRK2 kinase domain, due to their proximity to the ATP-binding pocket for a series of diaminopyrimidine compounds [132–135]. L1949 near the hinge region was targeted for selectivity since it is a smaller residue than Phe or Tyr, which is found in this position in approximately 60% of the kinome. Inhibitors with a nitrogen lone pair or substituents or that extend into this region can be accommodated with the Leu residue, but sterically clash with the larger Phe or Tyr. This strategy was used to obtain selectivity for LRRK2 over the JAK family of kinases and JAK2 was found to be a useful surrogate for predicting general kinome selectivity. Interactions with S1954 were targeted to achieve both general kinase selectivity and to overcome specific selectivity issues with the kinase TTK. Approximately 55% of the kinome has a larger group at this position, and approximately 50% has a negatively charged Asp or Glu residue in this position. Strategies to mitigate crossover to TTK, which has an Asp residue in this position, have included the introduction of small groups that would cause unfavorable steric and/or electrostatic interactions with the Asp side chain [132, 133, 135]. Selectivity of a series of 7-aryl-substituted quinoline derivatives may also be due to the placement of polar groups in the vicinity of the unconserved S1954 and R1957 residues [138]. Interactions with the unconserved R1957 and H1998 LRRK2 residues may also contribute to selectivity in a series of indolinone compounds [144]. Optimizing interactions with H1998 and the catalytic K1906 in this series of compounds was hypothesized to improve kinase selectivity, especially over RET kinase.

Alternative strategies used to obtain selectivity in kinases are through Type II inhibitors, which stabilize an inactive DFG-out conformation, or Type III inhibitors, which are not ATP-competitive. Only a handful of LRRK2 inhibitors have been reported that fall into these categories [131, 148]. These inhibitors were shown to be more potent against WT LRRK2 kinase compared to G2019S, supporting the hypothesis that the pathogenic G2019S mutation stabilizes the active conformation of LRRK2 [60, 114, 131, 136, 141]. This suggests a Type II kinase inhibitor may not be the optimal approach for the treatment of PD due to the LRRK2 G2019S mutation. The X-ray crystal structure of Roco4 kinase domain co-crystallized with the inhibitor H1152 revealed two binding sites for this inhibitor: one in the expected ATP-binding site and the other close to the αC-helix [118]. The implications of this second binding site for LRRK2 are unclear, but suggest not completely ruling out the possibility of developing Type III inhibitors for LRRK2 kinase.

4 Medicinal Chemistry: LRRK2 Kinase Inhibition

Historically, kinase inhibitors have been a key target for oncology indications. The critical role of kinases in cell cycle or apoptotic signaling pathways made them a logical target. From a drug discovery perspective for non-oncology indications, pharmacologically promiscuous kinase inhibitors can potentially give rise to safety

Table 2 Physicochemical property comparison of CNS marketed drugs and Pfizer CNS clinical candidates (CANs) with marketed kinase drugs

	MW	clogP	tPSA	clogD	HBD	pK_a	CNS MPO
CNS drugs[a] (121 cmpds)	298.7	2.72	47.6	1.70	1	6.2	4.61
CNS CANs[a] (108 cmpds)	357.4	3.44	53.6	2.16	1	7.5	4.25
Kinases[a] (21 cmpds)	461.2	4.24	85.6	2.63	2	5.6	3.56

[a]Average values

concerns due to their varied roles in signaling pathways. These safety issues are particularly acute for the treatment of chronic diseases that require long-term treatment, typically experienced with neurodegenerative disorders. Recently, kinase inhibitors have been developed for non-oncology indications. Xeljanz® (tofacitinib), a JAK3 inhibitor, is a good example and was approved in 2012 for moderate to severe rheumatoid arthritis (http://www.xeljanz.com/).

For a central nervous system (CNS) indication, in addition to the off-target selectivity, one needs to contend with the blood–brain barrier (BBB) and designing compounds that have access to the central compartment. The medicinal chemistry challenge is readily apparent when one analyzes the current marketed kinase chemical space relative to the historical CNS chemical space. From a kinase inhibitor perspective, the potential pharmaceutical agent must contend with equilibria of active, inactive, and intermediate conformations (e.g., DFG-in vs. DFG-out). Most of the initially marketed kinases targeted the DFG-out conformation which generates a much larger active site pocket, thus giving rise to compounds with MW, logP, or PSA parameters outside the range of typical CNS space. Comparison of these physicochemical parameters is presented in Table 2. These data were generated from an analysis of marketed kinases relative to the marketed CNS drugs and a set of preclinical CNS candidates from the Pfizer pipeline. Additionally, employing the CNS multiparameter optimization (MPO) score [149], as a guide to the likelihood the compound may be CNS drug like, clearly shows the current chemical matter is outside this space. Furthermore, as a consequence to the high concentration of ATP present in the cell, any potential kinase inhibitor will need to have high potency to be able to compete at a mass balance level.

For a CNS indication, such as PD, clearly having free drug in the brain at the site of action is an absolute requirement. Moreover, to minimize the potential for off-target toxicology, the requisite concentration of free drug in the brain should be achieved with minimal exposure in the periphery. Combining all these attributes generates a CNS kinase design strategy that incorporates the selection of chemical matter that has low MW and is polar and neutral with high brain availability. These compounds additionally will need to have high ligand binding efficiencies (LE), lipophilic efficiencies (LipE), and low efficacious concentration (C_{eff}), all the while having high kinome selectivity.

The current public domain LRRK2 kinase inhibitor chemical matter will be summarized with an eye to how they compare to the design strategy articulated (vide supra).

4.1 LRRK2 Patent Space Analysis

As a target, LRRK2 was first identified in 2004 and its connection to PD was confirmed by genome-wide associate studies (GWAS) in 2010 (For a recent PD genetics review: [150]). This short-time frame coupled with the lack of its (patho) physiological role(s) (vide supra) has presented a challenge to the medicinal chemist as witnessed by a relatively small compound set compared to targets with a more rich medicinal chemistry history. Conversely, the highly conserved nature of the ATP-binding site of kinases does give rise to the great potential for crossing-over of chemical scaffolds from inhibitors designed for alternate kinases. Indeed, the first set of LRRK2 kinase inhibitors investigated were compounds repurposed from other kinase programs (vide infra). A handful of LRRK2 selective kinase inhibitors have been published in the primary literature; however, they are typically specific examples from a greater set of compounds claimed in the LRRK2 patent literature. The Markush structures for this set of LRRK2 kinase inhibitors are presented in Fig. 5. In general terms, the LRRK2 ATP-binding site can accommodate known core scaffolds that incorporate 1-, 2-, and potentially 3-point hinge-binding motifs. Addressing the safety, CNS design features, and chemical novelty will arise from how these scaffolds are adorned with requisite substituents.

The attributes that the medicinal chemist has the greatest design control over are the physicochemical properties. All subsequently measured compound properties are embedded once the structure has been fixed. Figure 6 provides an overview of the LRRK2 patented chemical matter as it relates to physicochemical space. The box plots provide a summary of each property showing the data distribution with the average (red line), the CNS drug set average (green line), and the marketed kinase set (black dashed line). MW shows a normal distribution with the average between those for the kinase and CNS drug sets. This tendency towards the CNS drug set is not observed for PSA with the LRRK2 and kinase sets being essentially the same. Measures of lipophilicity (clogP and logD) again show a normal distribution with the average closer to the CNS drug set. HB donor count (HBD) for the LRRK2 compounds will be skewed to the kinase set due to the required hinge interactions, and this was observed. The distribution for pK_a appears to be bimodal for the LRRK2 set, but the average tends closer to the CNS drug set.

Figure 7 provides an overview of the probability of in vivo toxicology findings (tox plot) [151]. The upper left quadrant (clog$P > 3$ and PSA < 75) was determined to have the greatest likelihood of in vivo safety findings. Conversely, the lower right quadrant (clog$P < 3$ and PSA > 75) was found to have the lowest probability of a safety finding. The pie charts are scaled to the number of compounds in each quadrant, and Fig. 7 is color coded for CNS MPO desirability. Clearly, most of the compounds reside in good CNS space (MPO > 4), but a significant number of these compounds reside in physicochemical space expected to give rise to safety findings.

Figure 8 illustrates the breakdown of chemical matter by the patent assignee. Again, most organizations are working in chemical space that is congruent with CNS drugs, and in several instances, the companies have leveraged existing chemotypes from marketed kinases (Fig. 5). John Hopkins University has patented the use of existing kinase inhibitors, e.g., staurosporine, damnacanthal, SP600125,

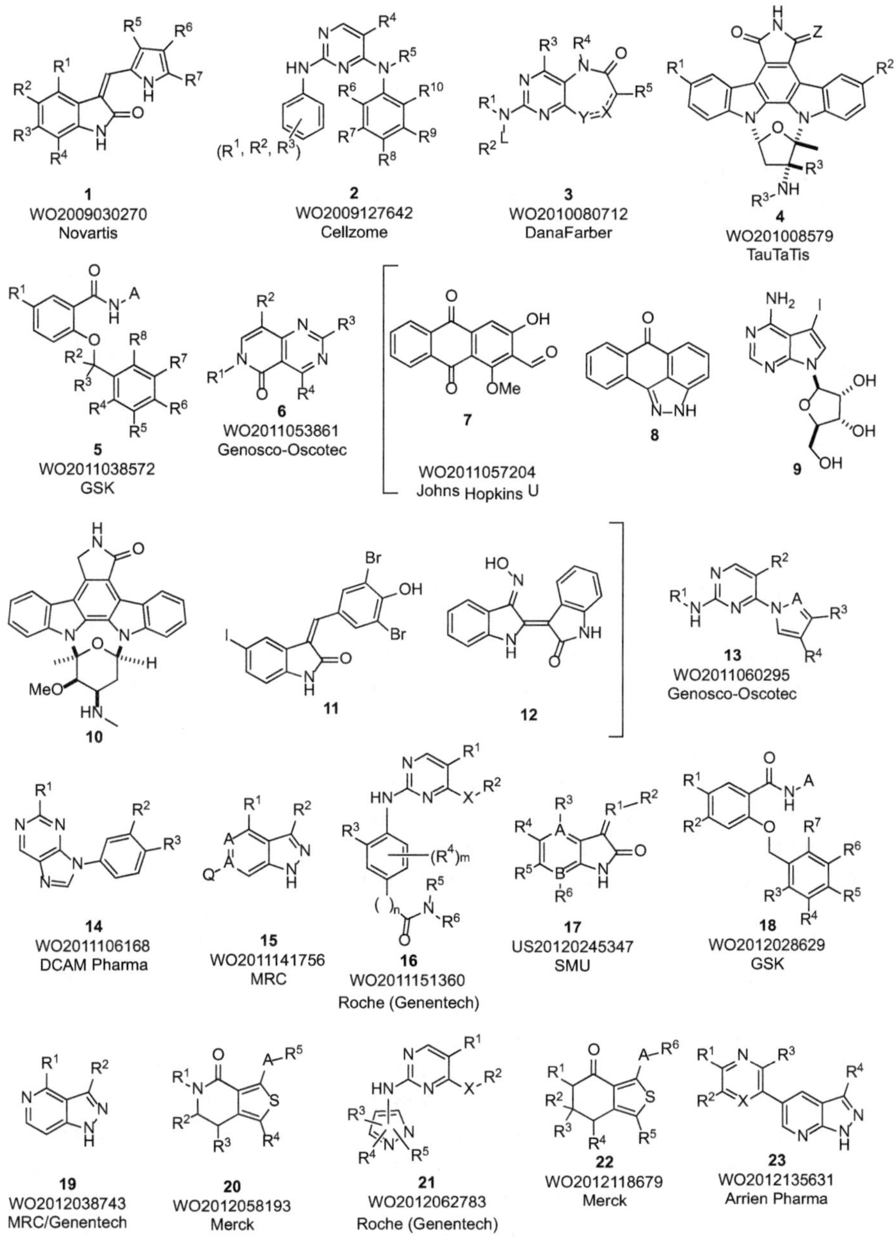

Fig. 5 (continued)

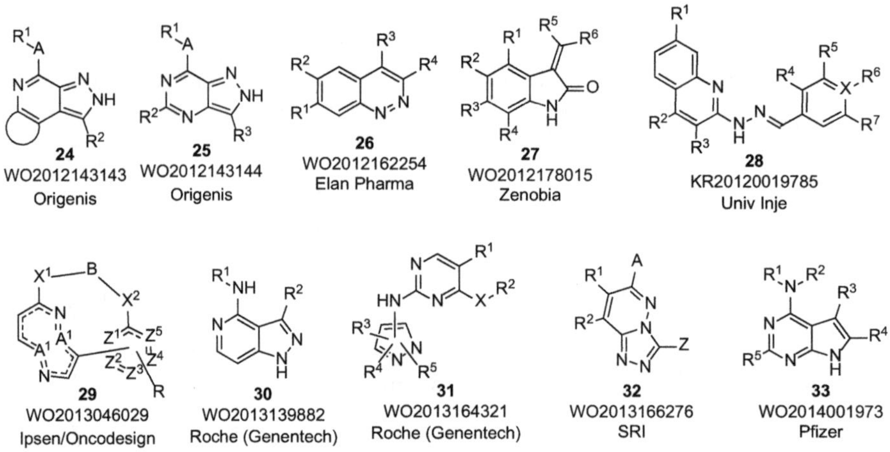

Fig. 5 LRRK2 kinase inhibitor Markush structures (chronological order)

5-iodotubercidin, GW5074, and indirubin-3′-monoxime as LRRK2 inhibitors, whereas Tautatis focused on modified versions of staurosporine analogues. Novartis, Southern Methodist University, and Zenobia focused on the oxindole framework as exemplified by sunitinib (Sutent®). The diaminopyrimidine core can be found in patents from Cellzome, Dana Farber, and Roche (Genentech). The Medical Research Council alone and in collaboration with Genentech, DCAM Pharma, Merck, Arrien, Origenis, Pfizer, and Southern Research Institute have all generated variations on a 6,5-fused bicyclic framework. Ipsen/Oncodesign extended the use of this scaffold by exemplifying macrocyclic variants of this core. GSK has disclosed a novel LRRK2 kinase inhibitor scaffold in the form of an aryl ether amide.

Comparing this set of compounds, as a whole, to the data for CNS drugs and marketed kinases (vide supra) provides potential insight as to the position in chemical space these compounds have relative to the space currently defined by the CNS drug and marketed kinase sets. All these compounds are ATP-competitive inhibitors presumably targeting the DYG-in conformation as the equilibrium distribution between the DYG-in and DYG-out (inactive conformation) appears to be skewed towards the active conformation by 5–6 kcal/mol [131]. Despite the lack of a LRRK2 crystal structure, one can leverage the knowledge from the kinase literature and predict that these compounds will interact with the hinge motif in a 1- or 2-point manner. Thus, to drive selectivity and ultimately safety, the substitution pattern from the core is crucial to accomplish this requisite attribute. From homology models, Genentech has identified two critical residues necessary to drive LRRK2 selectivity. These include the hinge residue Leu^{1949} and the activation loop residue Arg^{1957}. In the case of the diaminopyrimidines, Genentech was able to increase selectivity to off-targets (e.g., MST2 and JAK1) with substituents on the pendant aniline or aminopyrazole moieties.

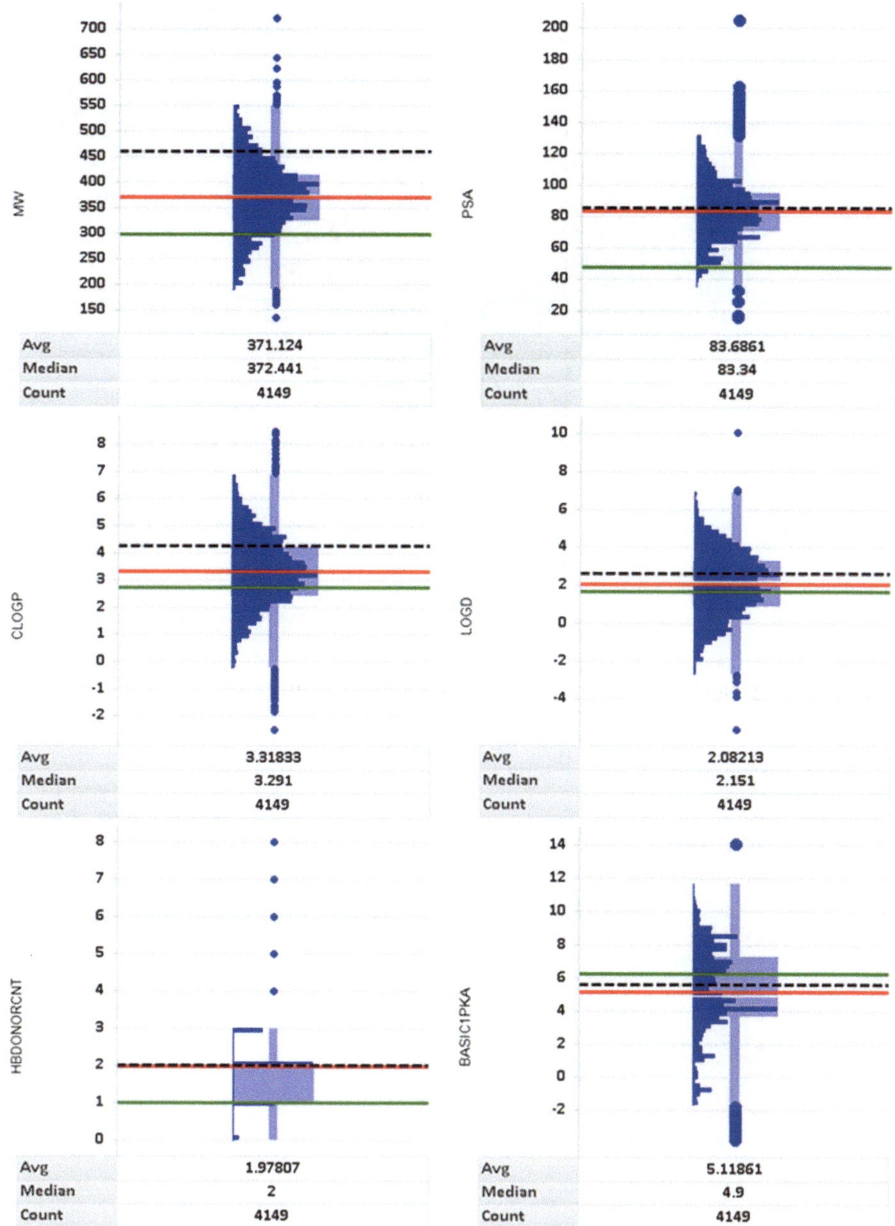

Fig. 6 Physicochemical analysis of LRRK2 patent chemical matter. Each box overlays the corresponding distribution for the descriptor on its box plots. The *dots* represent outliers and the *lines* denote LRRK2 data set average (*red*), CNS drug set average (*green*), and marketed kinase drugs average (*black dashed*)

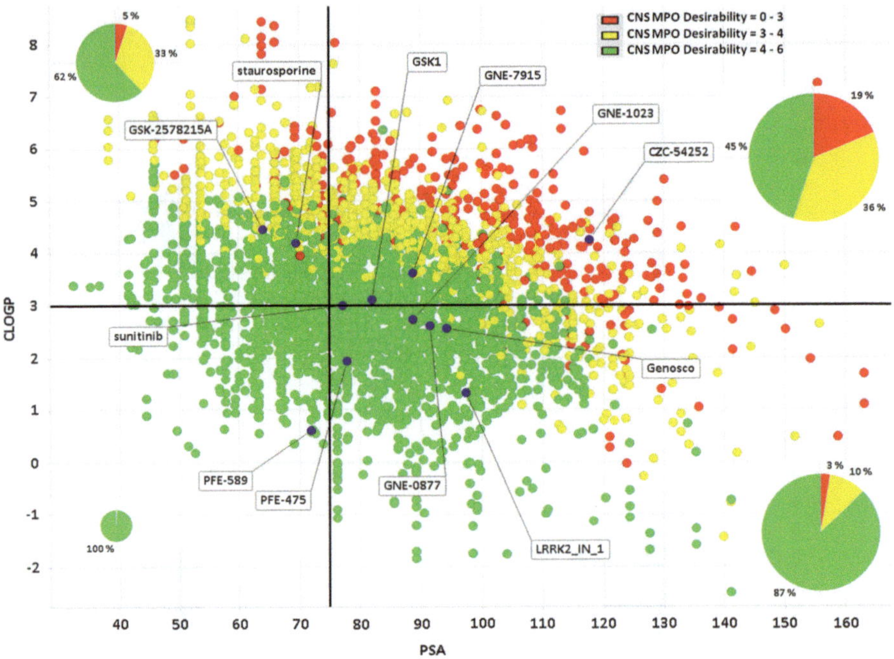

Fig. 7 Toxicity plot for LRRK2 patent chemical space (2009–2014)

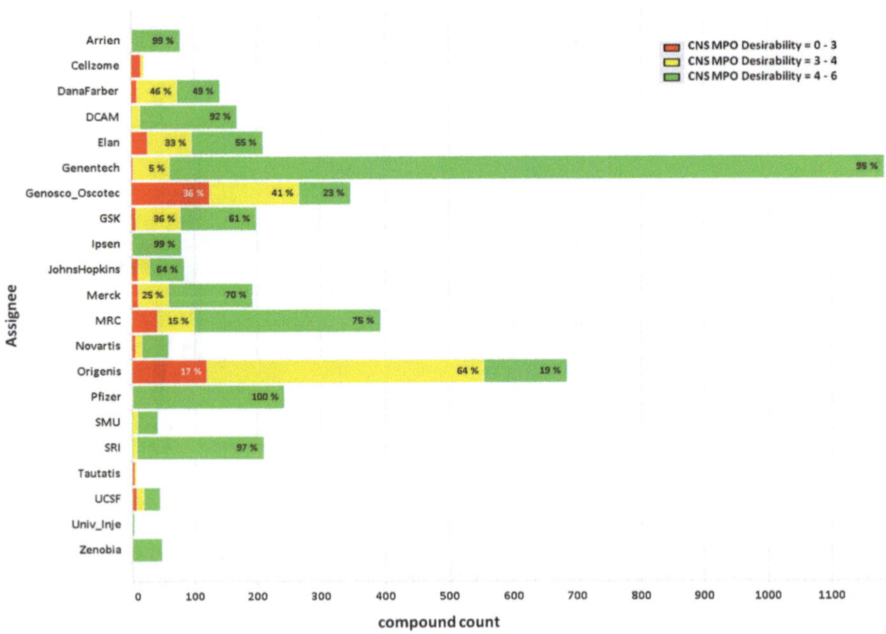

Fig. 8 CNS MPO desirability analysis for LRRK2 patents (2009–2014)

4.2 LRRK2 Chemical Matter Overview

It has been suggested that the ability to study a protein's function is enhanced when small molecule tools for the target exist [152]. Ideally a "compound toolbox" would be available, covering different chemical scaffolds and physicochemical and ADMET properties, and these compounds would be well characterized in terms of on- and off-target effects. With distinct scaffolds, overlap of off-target effects, for example, between the compounds is likely to be reduced, and consistent findings across the compounds would therefore be indicative of activity at the target in question.

The 28 patent applications of small molecules (Fig. 5), to date, represent roughly 10 unique chemical scaffolds. Through the use of SAR, X-ray crystal structures (surrogate crystallography), and homology models, specific residues in the LRRK2 ATP-binding pocket have been targeted for inhibitor interactions to improve potency and selectivity for LRRK2 kinase. Initial activities focused on LRRK2 potency and selectivity. As the compounds achieved these goals, focus shifted to address ADME and safety issues that were revealed as these compounds were used to expand LRRK2 biology. The diversity of the current chemical matter provides a toolbox of compounds that should enable the detailed elaboration of LRRK2 kinase function.

4.2.1 Repurposed Kinase Inhibitors

Early work in the field of LRRK2 kinase inhibitor development sought to compliment the range of biochemical tools that were becoming available [153] with small molecule kinase inhibitors. By screening panels of commercial compounds, groups began to identify known kinase inhibitors that also displayed activity against LRRK2 [35, 78, 154]. Initially identified compounds, using LanthaScreen detection and a surrogate peptide substrate (LRRKtide), included **7–9** and **11–12**. Staurosporine, **10**, was one of the most potent LRRK2 inhibitors reported (wt $IC_{50} = 6$ nM); however, it is also highly promiscuous across the kinome [155]. Capitalizing on this observation, **4** illustrates how modifications to the bis-indole core were pursued to potentially address the selectivity issue. Sunitinib, **34**, is a potent LRRK2 inhibitor (wt $IC_{50} = 37$ nM), but also displays similar promiscuity. Related indolidinones (**1**, **17**, and **27**) have been disclosed. Screening of a number of Rho-kinase inhibitors led to the identification of **35** [35]. Although it has a number of off-target effects, selectivity is much improved over **10** and **34**, although potency against LRRK2 is also reduced (wt $IC_{50} = 244$ nM).

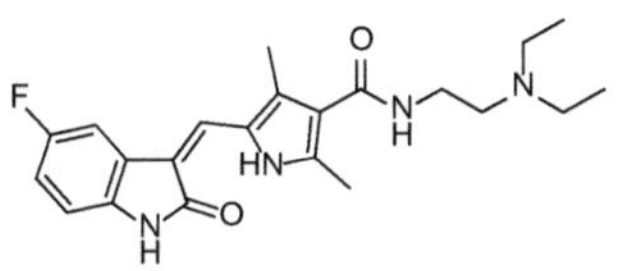

34

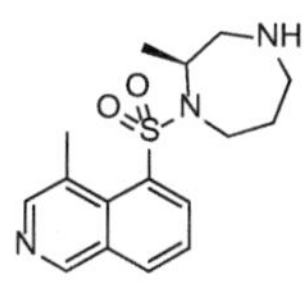

35

Although these early inhibitors are structurally different, they have in common potent activity across a range of kinases, and thus it is difficult to separate effects at LRRK2 from off-target effects, especially when examined in isolation. These compounds proved useful for initial study of the function of LRRK2; however, it was clear that the field could benefit from more selective compounds.

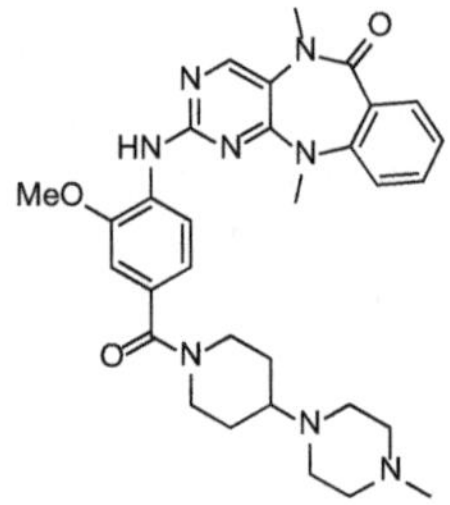

36

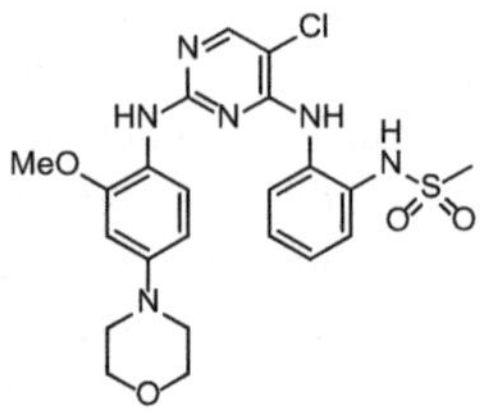

37

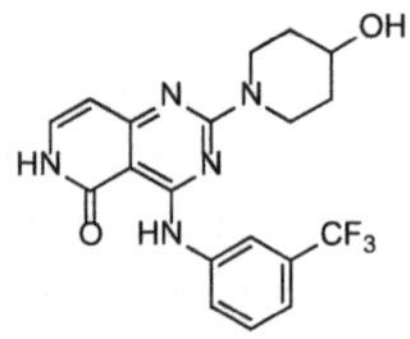

38

4.2.2 First-Generation LRRK2-Focused Kinase Inhibitors

In 2011, the first LRRK2-optimized kinase inhibitor LRRK2-IN-1, **36**, was reported (derived from **3**) [143]. Due to its high potency against both wild-type and mutant LRRK2 (IC_{50} values of 13 and 6 nM, respectively, measured in cell-free enzymatic assays), and its improved selectivity across a range of kinases, it was clearly a vast improvement on previous compounds and was initially employed throughout the field as a selective LRRK2 inhibitor. While **36** is potent against both LRRK2 WT and G2019S mutant enzymes, it has been demonstrated to display moderate potency in a whole cell assay, 200–600 nM [40, 146]. Further, high doses are required to observe any in vivo effects due to its low permeability and high levels of plasma protein binding. Additionally, no CNS effects were observed as **36** had low brain availability (unbound C_{brain}/unbound C_{plasma}). Although **36** demonstrated improved selectivity in comparison to previous LRRK2 inhibitors, it has some significant off-target potency, most notably displaying equipotent activity against ERK5 (BMK1, MAPK7), DCAMKL, PLK1, and PLK4. This off-target activity confounded some of the biological assays in which the effects of **36** are attributed to LRRK2, in particular when assessing effects on neurite outgrowth and inflammation end-points [40]; consequently, use of **36** has diminished.

4.2.3 Second-Generation LRRK2-Focused Kinase Inhibitors

As intensive research into the function and dysfunction of LRRK2 and its mutants continued, a number of other tool compounds and inhibitor series have been published from both academic and industrial laboratories. The most potent of these derived from **2**, in both enzyme (5–7 nM in WT and G2019S, respectively) and cell-based assays (attenuation of G2019S and R1441C-induced neuronal injury and death in a concentration-dependent manner with an EC_{50} of <10 nM), was CZC-54252 (**37**) reported in 2011 [156]. Unfortunately, **37** had disappointing selectivity across the kinome. Although CZC-54252 may prove a useful tool for cell-based assays, its poor CNS MPO score (2.12) suggested it was outside of CNS space which will limit its application in in vivo models (confirmed with a reported brain availability of 0.05).

The diaminopyrimidine core has been extensively examined by several groups. In addition to compounds derived from **2** and **3** (vide infra), variations including analogues from **6**, **13**, **16**, **21**, and **31** have been reported. An early example is exemplified by **38** (CNS MPO = 3.97), but more advanced analogues successfully balanced good physicochemical properties with high degrees of LRRK2 selectivity [157, 158]. Focusing on diaminopyrimidine HTS hits, due to high ligand efficiency and reasonable PK properties, optimization of the hits using a LRRK2 homology model to identify both key interactions for LRRK2 potency and sites that may offer enhanced kinase selectivity yielded **39** (MPO = 5.37) with improved kinase selectivity over their initial hit, suitable in vivo exposures, and similar cellular activity (IC_{50} = 389 nM) to **36**. While disclosed in a patent the previous year (WO2012062783; see Fig. 5, above), **39** was subsequently described as HG-10-102-01 and reported to exhibit in vivo mouse brain activity [132].

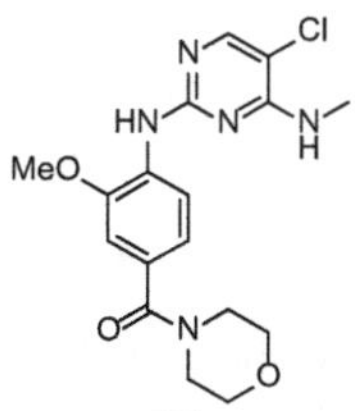

39

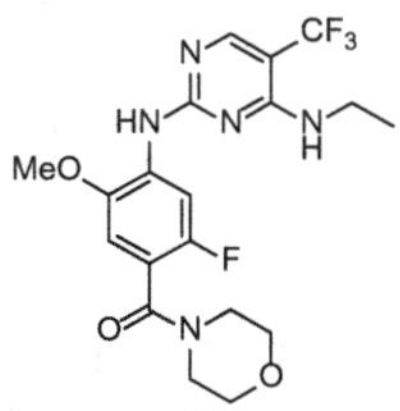

40

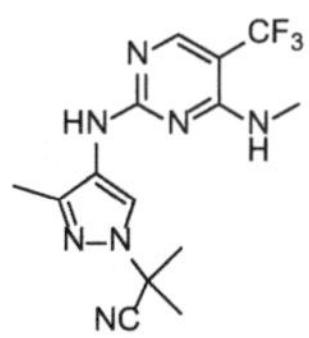

41

Having identified TTK (MPS1) as the major off-target interaction of this series, and significant safety concern, continued optimization by Genentech scientists, using structure-based drug design initially arrived at **40** (CNS MPO = 4.60), a compound having excellent selectivity for LRRK2 [145]. The in vivo activity of this compound and related analogues was shown using dephosphorylation of S1292 as a measure of LRRK2 kinase activity. Although **40** is an excellent tool compound for probing the effect of LRRK2 inhibition in vivo, it has some major liabilities, thus further optimization was needed for progression towards a potential clinical candidate. As the diaminopyrimidine forms the key hinge interactions with LRRK2, optimization was focused around the side chain. The aniline motif, which could lead to idiosyncratic toxicity findings, was replaced with an aminopyrazole, and side chains to enhance aqueous solubility were incorporated [133–135]. This optimization lead to GNE-0877, **41** (CNS MPO = 5.45), which has been progressed to preclinical safety studies (see Sect. 2.6, above).

4.2.4 Third-Generation LRRK2-Focused Kinase Inhibitors

Moving away from diaminopyrimidines, several other chemotypes have been reported. A series defined by **5** and **18**, while likely to be ATP-competitive, do not contain a common hinge-binding motif [135]. This attribute may be responsible for the excellent kinase selectivity. GSK2578215A, **42**, is a representative analogue from this series and showed reasonable potency ($IC_{50} = 47$ nM). However, it has poor rodent PK, with a relatively low oral bioavailability (%$F = 12$) and half life of 1.1 h [159]. The lack of LRRK2 inhibition observed in the brain (potentially predicted by its CNS MPO score of 3.82), in comparison to kidney and spleen, suggested low levels of unbound compound in brain potentially limiting the use of this series for in vivo models.

Quinoline derivatives, **26** and **28**, are known single-point hinge binders and have been found to be good LRRK2 kinase inhibitors [139, 147]. Although the cinnoline variants [141] were potent LRRK2 inhibitors (wt $IC_{50} = 7$ nM), they were found to be promiscuous in a small kinase panel. By focusing the core to quinoline [138], e.g., **43**, kinase selectivity was significantly improved and in vivo activity was demonstrated (CNS MPO = 5.05).

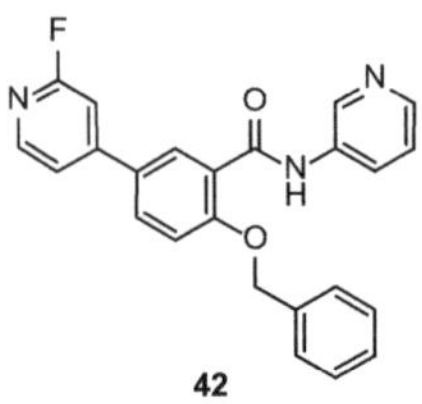

42

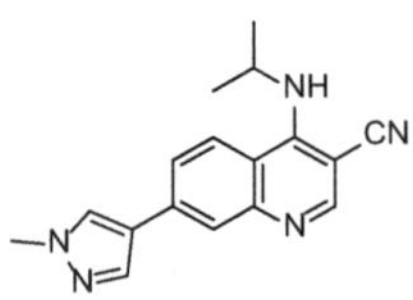

43

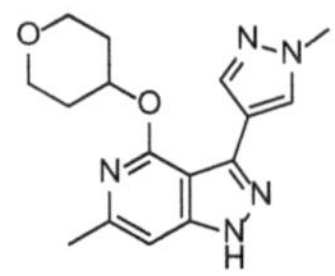

44

Pyrazolopyri(mi)dines, **15**, **19**, **23–25**, and **30**, have been well represented in the patent literature. For analogues, such as **44** (CNS MPO = 5.83), potency is comparable to **36** with overall good ADME [159]. Alternate 6,5-fused ring systems, **14**, **20**, and **22**, have been reported; however, the pyrrolopyrimidine **33** appears to have the best alignment of properties. Specifically, **45** (CNS MPO = 5.83) is one of the most active LRRK2 inhibitors in vivo (brain free drug wt $IC_{50} = 15$ nM) along with high kinase selectivity [160].

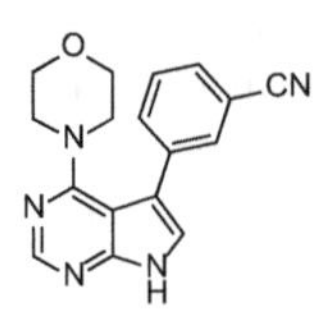

45

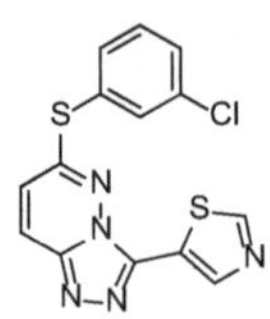

46

One final variation of the 6,5-fused ring system is illustrated by **32**. These compounds are also postulated to be single-point hinge binders, interacting through one of the triazole nitrogens. A representative compound from this series, **46** (CNS MPO = 4.88), displayed excellent in vitro potency (wt $IC_{50} = 31$ nM; G2019S $IC_{50} = 8$ nM) against LRRK2, but this did not translate to cellular assays resulting in a 100-fold right shift [137]. As witnessed with other scaffolds (vide supra), compounds from this chemotype would make good tool compounds for probing LRRK2 in vitro biology.

4.2.5 Pharmacological Profiles for Key LRRK2-Focused Kinase Inhibitors

As the relatively young LRRK2 field of research has evolved, there have been frequent advances in the biochemical reagents available, as the biology has developed. In part, this has been facilitated by the improvement of the chemical tools (vide supra). However, the literature data documenting the pharmacological profiles of the tool compounds is often not consistent in terms of the assays or conditions used in reporting their potencies. Thus, Table 3 presents a head-to-head comparison of several of the top compounds, from our internal assessments, with the intention of providing LRRK2 researchers with additional data to better enable selecting of the best compound(s) for their studies.

The LRRK2 inhibitors described below have been characterized in a number of in vitro and in vivo models. Activity against LRRK2 wild-type and G2019S mutant was assayed using truncated enzyme from Invitrogen in a LanthaScreen format at 50 μM and 1 mM ATP concentrations [58]. The whole cell potency is measured in HEK293 cells transfected with WT LRRK2, using the antibody against pS935 [64]. The RRCK (Ralph Russ Canine Kidney) assay is a measure of passive permeability, using a canine kidney cell line which expresses low levels of efflux transporters [161]. MDR1 is a measure of efflux by human P-gp transporter and is used to assess the likelihood of brain impairment [162]. HLM and RLM are an assessment of metabolism in human and rat liver microsomes, respectively. Several safety assays are performed; "dof" is a measure of the ability of compounds to displace [^{3}H]-labeled dofetilide from the hERG K^+-channel, important for cardiac safety in clinical candidates [163]. General cytotoxicity is measured using the THLE assay [164]. Effects on cell health are followed up with a panel of assays performed in HepG2 cells, measuring mitochondrial function [165]. Kinome selectivity is reported from a Dundee Panel [166] using recombinant enzymes at the ATP K_M or ActivX data [167], which used lysed PBMC cells to assay selectivity in a more native system.

Table 3 Pharmacological profile of key LRRK2 kinase inhibitors

	Assay	36	37	42	40	45
LRRK2 (in vitro)	1 mM ATP LRRK2 IC_{50} (nM)	68	22	150	37	8
	1 mM ATP LRRK2 (G2019S) IC_{50} (nM)	25	13	247	23	34
	pS935 WCA LRRK2 IC_{50} (nM)	233	4.95	992	194	53
	ActivX huPBMC LRRK2 IC_{50} (nM)	> 300 [142]	15	400	nt	15
Physico-chem	LE	0.23	0.31	0.31	0.33	0.48
	LipE	5	4.1	2.5	3.5	5.5
	CNS MPO	4.52	2.12	3.82	4.6	5.83
ADME	RRCK AB (x 10^{-6} cm/sec)	3.88	14.7	2.87	12.7	27.2
	MDR1 ER	22	3.67	2.24	1.07	1.02
	HLM $CL_{IA,S}$ (mL/min/kg)	22	17	< 8	< 8	36
	RLM $CL_{IA,S}$ (mL/min/kg)	39	101	nt	nt	89
Safety	Dof IC_{50} (μM)	32	16	> 100	50	44
	HepG2 (24h) Gal IC_{50} (μM)	90	> 300	> 300	103	> 300
	HepG2 (24h) Glu IC_{50} (μM)	103	> 300	> 300	159	> 300
	HepG2 (72h) Glu IC_{50} (μM)	26	112	> 300	42	> 300
	THLE Cv IC_{50} (μM)	18	17	> 300	75	> 223
Rat BA	C_{bu}/C_{pu} (AUC)	< 0.1	nt	nt	0.5	1
KSS	Dundee Panel 1 μM (hits\total)	2/121	15/131	2/131	20/140	7/159[a]
LRRK2 (*in vivo*)	Mouse brain free drug pS935 IC_{50} (nM)	7,798[b]	nt	nt	75	<20

Colors represent a stop-light analysis of the data based on Pfizer cut-offs (green low; yellow moderate; red high)
[a] ActivX data at 1 μM
[b] Kidney data. *nt*, not tested. See text for description of assays

5 Summary, Conclusions, and Outlook

The answers to some key questions are important for a more thorough understanding of LRRK2 structure, function, and the development of inhibitors, including how the Roc domain regulates the kinase domain, what the role of the COR domain is, and how the pathogenic mutations alter the function of the LRRK2 protein. X-ray crystal structures of the LRRK2 enzymatic domains, including both WT and the pathogenic mutations, could contribute to an understanding of these questions and would be a key accomplishment to help drive inhibitor design.

The strength of research in the field of LRRK2, both academic and industrial, has generated a wealth of chemical tools for the study of LRRK2. Where the initially reported compounds had activity across multiple kinase targets, new inhibitors have a high degree of kinome selectivity along with improved physicochemical properties, providing opportunities to study inhibition of LRRK2 kinase activity from the isolated enzyme through to in vivo assays. Ultimately, this collection of inhibitors should allow researchers the flexibility to design experiments to probe LRRK2 function, knowing that they have the right tools for the job.

References

1. Antony PM, Diederich NJ, Krüger R, Balling R (2013) The hallmarks of Parkinson's disease. FEBS J 280:5981
2. Deleidi M, Gasser T (2013) The role of inflammation in sporadic and familial Parkinson's disease. Cell Mol Life Sci 70:4259
3. Pradhan S, Andreasson K (2013) Commentary: progressive inflammation as a contributing factor to early development of Parkinson's disease. Exp Neurol 241:148
4. Reeve A, Simcox E, Turnbull D (2014) Ageing and Parkinson's disease: why is advancing age the biggest risk factor? Ageing Res Rev 14:19
5. Singleton AB, Farrer MJ, Bonifati V (2013) The genetics of Parkinson's disease: progress and therapeutic implications. Mov Disord 28:14
6. Paisan-Ruiz C, Jain S, Evans EW et al (2004) Cloning of the gene containing mutations that cause PARK8-linked Parkinson's disease. Neuron 44:595
7. Zimprich A, Biskup S, Leitner P et al (2004) Mutations in LRRK2 cause autosomal-dominant parkinsonism with pleomorphic pathology. Neuron 44:601
8. Healy DG, Falchi M, O'Sullivan S, Bonifati V, Durr A, Bressman S, Brice A, Aasly J, Zabetian CP, Goldwurm S, Ferreira JJ, Tolosa E, Kay DM, Klein C, Williams DR, Marras C, Lang AE, Wszolek ZW, Bericiano J, Schapira AHV, Lynch T, Bhatia KP, Gasser T, Lees AJ, Wood NW (2008) Phenotype, genotype, and worldwide genetic penetrance of LRRK2-associated Parkinson's disease: a case–control study. Lancet Neurol 7:583
9. Goldwurm S, Zini M, Mariani L et al (2007) Evaluation of LRRK2 G2019S penetrance: relevance for genetic counseling in Parkinson disease. Neurology 68:1141
10. Latourelle JC, Sun M, Lew MF et al (2008) The Gly2019Ser mutation in LRRK2 is not fully penetrant in familial Parkinson's disease: the GenePD study. BMC Med 6:32
11. Gasser T (2009) Molecular pathogenesis of Parkinson disease: insights from genetic studies. Expert Rev Mol Med 11:e22
12. Aasly JO, Vilariño-Güell C, Dachsel JC, Webber PJ, West AB, Haugarvoll K, Johansen KK, Toft M, Nutt JG, Payami H, Kachergus JM, Lincoln SJ, Felic A, Wider C, Soto-Ortolaza AI, Cobb SA, White LR, Ross OA, Farrer MJ (2010) Novel pathogenic LRRK2 p. Asn1437His substitution in familial Parkinson's disease. Mov Disord 25:2156
13. Lorenzo-Betancor O, Samaranch L, Ezquerra M, Tolosa E, Lorenzo E, Irigoyen J, Gaig C, Paster MA, Soto-Ortolaza AI, Ross OA, Rodriguez-Oroz MC, Valldeoriola F, Marti MJ, Luquin MR, Perez-Tur J, Burguera JA, Obeso JA, Pastor P (2012) LRRK2 haplotype-sharing analysis in Parkinson's disease reveals a novel p.S1761R mutation. Mov Disord 27:146
14. Taymans JM, Vancraenenbroeck R, Ollikainen P, Beilina A, Lobbestael E, De Maeyer M, Baekelandt V, Cookson MR (2011) LRRK2 kinase activity is dependent on LRRK2 GTP binding capacity but independent of LRRK2 GTP binding. PLoS One 6:e23207
15. Biosa A, Trancikova A, Civiero L, Glauser L, Bubacco L, Greggio E, Moore DJ (2013) GTPase activity regulates kinase activity and cellular phenotypes of Parkinson's disease-associated LRRK2. Hum Mol Genet 22:1140
16. Sheng Z, Zhang S, Bustos D, Kleinheinz T, Le Pichon CE, Dominguez SL, Solanoy HO, Drummond J, Zhang X, Ding X, Cai F, Song Q, Li X, Yue Z, van der Brug MP, Burdick DJ, Gunzner-Toste J, Chen H, Liu X, Estrada AA, Sweeney ZK, Scearce-Levie K, Moffat JG, Kirkpatrick DS, Zhu H (2012) Ser1292 autophosphorylation is an indicator of LRRK2 kinase activity and contributes to the cellular effects of PD mutations. Sci Transl Med 4:164ra161

17. Haugarvoll K, Rademakers R, Kachergus JM, Nuytemans K, Ross OA, Gibson JM, Tan EK, Gaig C, Tolosa E, Goldwurm S, Guidi M, Riboldazzi G, Brown L, Walter U, Benecke R, Berg D, Gasser T, Theuns J, Pals P, Cras P, De Deyn PP, Engelborghs S, Pickut B, Uitti RJ, Foroud T, Nichols WC, Hagenah J, Klein C, Samii A, Zabetian CP, Bonifati V, Van Broeckhoven C, Farrer MJ, Wszolek ZK (2008) LRRK2 R1441C parkinsonism is clinically similar to sporadic Parkinson disease. Neurology 70:1456
18. Hyun CH, Yoon CY, Lee HJ, Lee SJ (2013) LRRK2 as a potential genetic modifier of synucleinopathies: interlacing the two major genetic factors of Parkinson's disease. Exp Neurobiol 22:249
19. Russo I, Bubacco L, Greggio E (2014) LRRK2 and neuroinflammation: partners in crime in Parkinson's disease? J Neuroinflammation 11:52
20. Zuo L, Motherwell MS (2013) The impact of reactive oxygen species and genetic mitochondrial mutations in Parkinson's disease. Gene 532:18
21. Manzoni C, Lewis PA (2013) Dysfunction of the autophagy/lysosomal degradation pathway is a shared feature of the genetic synucleinopathies. FASEB J 27:3424
22. Gardet A, Benita Y, Li C, Sands BE, Ballester I, Stevens C, Korzenik JR, Rioux JD, Daly MJ, Xavier RJ, Podolsky DK (2010) LRRK2 is involved in the IFN-gamma response and host response to pathogens. J Immunol 185:5577
23. Moehle MS, Webber PJ, Tse T, Sukar N, Standaert DG, DeSilva TM, Cowell RM, West AB (2012) LRRK2 inhibition attenuates microglial inflammatory responses. J Neurosci 32:1602
24. Cho HJ, Liu G, Jin SM, Parisiadou L, Xie C, Yu J, Sun L, Ma B, Ding J, Vancraenenbroeck R, Lobbestael E, Baekelandt V, Taymans JM, He P, Troncoso JC, Shen Y, Cai H (2013) MicroRNA-205 regulates the expression of Parkinson's disease-related leucine-rich repeat kinase 2 protein. Hum Mol Genet 22:608
25. Guerreiro PS, Huang Y, Gysbers A, Cheng D, Gai WP, Outeiro TF, Halliday GM (2013) LRRK2 interactions with α-synuclein in Parkinson's disease brains and in cell models. J Mol Med 91:513
26. Greggio E, Zambrano I, Kaganovich A, Beilina A, Taymans JM, Daniëls V, Lewis P, Jain S, Ding J, Syed A, Thomas KJ, Baekelandt V, Cookson MR (2008) The Parkinson disease-associated leucine-rich repeat kinase 2 (LRRK2) is a dimer that undergoes intramolecular autophosphorylation. J Biol Chem 283:16906
27. Sen S, Webber PJ, West AB (2009) Dependence of leucine-rich repeat kinase 2 (LRRK2) kinase activity on dimerization. J Biol Chem 284:36346
28. Berger Z, Smith KA, Lavoie MJ (2010) Membrane localization of LRRK2 is associated with increased formation of the highly active LRRK2 dimer and changes in its phosphorylation. Biochemistry 49:5511
29. Civiero L, Vancraenenbroeck R, Belluzzi E, Beilina A, Lobbestael E, Reyniers L, Gao F, Micetic I, De Maeyer M, Bubacco L, Baekelandt V, Cookson MR, Greggio E, Taymans JM (2012) Biochemical characterization of highly purified leucine-rich repeat kinases 1 and 2 demonstrates formation of homodimers. PLoS One 7:e43472
30. Webber PJ, Smith AD, Sen S, Renfrow MB, Mobley JA, West AB (2011) Autophosphorylation in the leucine-rich repeat kinase 2 (LRRK2) GTPase domain modifies kinase and GTP-binding activities. J Mol Biol 412:94
31. James NG, Digman MA, Gratton E, Barylko B, Ding X, Albanesi JP, Goldberg MS, Jameson DM (2012) Number and brightness analysis of LRRK2 oligomerization in live cells. Biophys J 102:L41
32. Schapansky J, Nardozzi JD, Felizia F, Lavoie MJ (2014) Membrane recruitment of endogenous LRRK2 precedes its potent regulation of autophagy. Hum Mol Genet. doi:10.1093/hmg/ddu138
33. Davies P, Hinkle KM, Sukar NN, Sepulveda B, Mesias R, Serrano G, Alessi DR, Beach TG, Benson DL, White CL, Cowell RM, Das SS, West AB, Melrose HL (2013) Comprehensive characterization and optimization of leucine rich repeat kinase 2 (LRRK2) monoclonal antibodies. Biochem J 453:101

34. Jaleel M, Nichols RJ, Deak M, Campbell DG, Gillardon F, Knebel A, Alessi DR (2007) LRRK2 phosphorylates moesin at threonine-558: characterization of how Parkinson's disease mutants affect kinase activity. Biochem J 405:307
35. Nichols RJ, Dzamko N, Hutti JE, Cantley LC, Deak M, Moran J, Bamborough P, Reith AD, Alessi DR (2009) Substrate specificity and inhibitors of LRRK2, a protein kinase mutated in Parkinson's disease. Biochem J 424:47
36. Imai Y, Gehrke S, Wang HQ, Takahashi R, Hasegawa K, Oota E, Lu B (2008) Phosphorylation of 4E-BP by LRRK2 affects the maintenance of dopaminergic neurons in Drosophila. EMBO J 27:2432
37. Kumar A, Greggio E, Beilina A, Kaganovich A, Chan D, Taymans JM, Wolozin B, Cookson MR (2010) The Parkinson's disease associated LRRK2 exhibits weaker in vitro phosphorylation of 4E-BP compared to autophosphorylation. PLoS One 5:e8730
38. Trancikova A, Mamais A, Webber PJ, Stafa K, Tsika E, Glauser L, West AB, Bandopadhyay R, Moore DJ (2012) Phosphorylation of 4E-BP1 in the mammalian brain is not altered by LRRK2 expression or pathogenic mutations. PLoS One 7:e47784
39. Qing H, Wong W, McGeer EG, McGeer PL (2009) Lrrk2 phosphorylates alpha synuclein at serine 129: Parkinson disease implications. Biochem Biophys Res Commun 387:149
40. Luerman GC, Nguyen C, Samaroo H, Loos P, Xi H, Hurtado-Lorenzo A, Needle E, Stephen Noell G, Galatsis P, Dunlop J, Geoghegan KF, Hirst WD (2014) Phosphoproteomic evaluation of pharmacological inhibition of leucine-rich repeat kinase 2 reveals significant off-target effects of LRRK-2-IN-1. J Neurochem 128:561
41. Ohta E, Kawakami F, Kubo M, Obata F (2011) LRRK2 directly phosphorylates Akt1 as a possible physiological substrate: impairment of the kinase activity by Parkinson's disease-associated mutations. FEBS Lett 585:2165
42. Stafa K, Trancikova A, Webber PJ, Glauser L, West AB, Moore DJ (2012) GTPase activity and neuronal toxicity of Parkinson's disease-associated LRRK2 is regulated by ArfGAP1. PLoS Genet 8:e1002526
43. Xiong Y, Yuan C, Chen R, Dawson TM, Dawson VL (2012) ArfGAP1 is a GTPase activating protein for LRRK2: reciprocal regulation of ArfGAP1 by LRRK2. J Neurosci 32:3877
44. Gillardon F (2009) Leucine-rich repeat kinase 2 phosphorylates brain tubulin-beta isoforms and modulates microtubule stability–a point of convergence in Parkinsonian neurodegeneration? J Neurochem 110:1514
45. Matta S, Van Kolen K, da Cunha R, van den Bogaart G, Mandemakers W, Miskiewicz K, De Bock PJ, Morais VA, Vilain S, Haddad D, Delbroek L, Swerts J, Chávez-Gutiérrez L, Esposito G, Daneels G, Karran E, Holt M, Gevaert K, Moechars DW, De Strooper B, Verstreken P (2012) LRRK2 controls an EndoA phosphorylation cycle in synaptic endocytosis. Neuron 75:1008
46. Parisiadou L, Xie C, Cho HJ, Lin X, Gu XL, Long CX, Lobbestael E, Baekelandt V, Taymans JM, Sun L, Cai H (2009) Phosphorylation of ezrin/radixin/moesin proteins by LRRK2 promotes the rearrangement of actin cytoskeleton in neuronal morphogenesis. J Neurosci 29:13971
47. Kanao T, Venderova K, Park DS, Unterman T, Lu B, Imai Y (2010) Activation of FoxO by LRRK2 induces expression of proapoptotic proteins and alters survival of postmitotic dopaminergic neuron in Drosophila. Hum Mol Genet 19:3747
48. Pungaliya PP, Bai Y, Lipinski K, Anand VS, Sen S, Brown EL, Bates B, Reinhart PH, West AB, Hirst WD, Braithwaite SP (2010) Identification and characterization of a leucine-rich repeat kinase 2 (LRRK2) consensus phosphorylation motif. PLoS One 5:e13672
49. Zach S, Felk S, Gillardon F (2010) Signal transduction protein array analysis links LRRK2 to Ste20 kinases and PKC zeta that modulate neuronal plasticity. PLoS One 5:e13191
50. Gloeckner CJ, Schumacher A, Boldt K, Ueffing M (2009) The Parkinson disease-associated protein kinase LRRK2 exhibits MAPKKK activity and phosphorylates MKK3/6 and MKK4/7, in vitro. J Neurochem 109:959

51. Chen CY, Weng YH, Chien KY, Lin KJ, Yeh TH, Cheng YP, Lu CS, Wang HL (2012) (G2019S) LRRK2 activates MKK4-JNK pathway and causes degeneration of SN dopaminergic neurons in a transgenic mouse model of PD. Cell Death Differ 19:1623
52. Hsu CH, Chan D, Greggio E, Saha S, Guillily MD, Ferree A, Raghavan K, Shen GC, Segal L, Ryu H, Cookson MR, Wolozin B (2010) MKK6 binds and regulates expression of Parkinson's disease-related protein LRRK2. J Neurochem 112:1593
53. Martin I, Kim JW, Lee BD, Kang HC, Xu JC, Jia H, Stankowski J, Kim MS, Zhong J, Kumar M, Andrabi SA, Xiong Y, Dickson DW, Wszolek ZK, Pandey A, Dawson TM, Dawson VL (2014) Ribosomal protein s15 phosphorylation mediates LRRK2 neurodegeneration in Parkinson's disease. Cell 157:472
54. Yun HJ, Park J, Ho DH, Kim H, Kim CH, Oh H, Ga I, Seo H, Chang S, Son I, Seol W (2013) LRRK2 phosphorylates Snapin and inhibits interaction of Snapin with SNAP-25. Exp Mol Med 45:e36
55. Kawakami F, Yabata T, Ohta E, Maekawa T, Shimada N, Suzuki M, Maruyama H, Ichikawa T, Obata F (2012) LRRK2 phosphorylates tubulin-associated tau but not the free molecule: LRRK2-mediated regulation of the tau-tubulin association and neurite outgrowth. PLoS One 1:e30834
56. Bailey RM, Covy JP, Melrose HL, Rousseau L, Watkinson R, Knight J, Miles S, Farrer MJ, Dickson DW, Giasson BI, Lewis J (2013) LRRK2 phosphorylates novel tau epitopes and promotes tauopathy. Acta Neuropathol 126:809
57. West AB, Moore DJ, Biskup S, Bugayenko A, Smith WW, Ross CA, Dawson VL, Dawson TM (2005) Parkinson's disease-associated mutations in leucine-rich repeat kinase 2 augment kinase activity. Proc Natl Acad Sci U S A 102:16842
58. Anand VS, Reichling LJ, Lipinski K, Stochaj W, Duan W, Kelleher K, Pungaliya P, Brown EL, Reinhart PH, Somberg R, Hirst WD, Riddle SM, Braithwaite SP (2009) Investigation of leucine-rich repeat kinase 2: enzymological properties and novel assays. FEBS J 276:466
59. Gloeckner CJ, Kinkl N, Schumacher A, Braun RJ, O'Neill E, Meitinger T, Kolch W, Prokisch H, Ueffing M (2006) The Parkinson disease causing LRRK2 mutation I2020T is associated with increased kinase activity. Hum Mol Genet 15:223
60. Ray S, Bender S, Kang S, Lin R, Glicksman MA, Liu M (2014) The Parkinson disease-linked LRRK2 protein mutation I2020T stabilizes an active state conformation leading to increased kinase activity. J Biol Chem 289:13042
61. West AB, Moore DJ, Choi C, Andrabi SA, Li X, Dikeman D, Biskup S, Zhang Z, Lim KL, Dawson VL, Dawson TM (2007) Parkinson's disease-associated mutations in LRRK2 link enhanced GTP-binding and kinase activities to neuronal toxicity. Hum Mol Gene 16:223
62. Gloeckner CJ, Boldt K, von Zweydorf F, Helm S, Wiesent L, Sarioglu H, Ueffing M (2010) Phosphopeptide analysis reveals two discrete clusters of phosphorylation in the N-terminus and the Roc domain of the Parkinson-disease associated protein kinase LRRK2. J Proteome Res 9:173
63. Kamikawaji S, Ito G, Sano T, Iwatsubo T (2013) Differential effects of familial parkinson mutations in LRRK2 revealed by a systematic analysis of autophosphorylation. Biochemistry 52:6052
64. Dzamko N, Deak M, Hentati F, Reith AD, Prescott AR, Alessi DR, Nichols RJ (2010) Inhibition of LRRK2 kinase activity leads to dephosphorylation of Ser(910)/Ser(935), disruption of 14-3-3 binding and altered cytoplasmic localization. Biochem J 430:405
65. Nichols RJ, Dzamko N, Morrice NA, Campbell DG, Deak M, Ordureau A, Macartney T, Tong Y, Shen J, Prescott AR, Alessi DR (2010) 14-3-3 binding to LRRK2 is disrupted by multiple Parkinson's disease-associated mutations and regulates cytoplasmic localization. Biochem J 430:393
66. Gillardon F, Kremmer E, Froehlich T, Ueffing M, Hengerer B, Gloeckner CJ (2013) ATP-competitive LRRK2 inhibitors interfere with monoclonal antibody binding to the kinase domain of LRRK2 under native conditions. A method to directly monitor the active conformation of LRRK2? J Neurosci Methods 214:62

67. Beilina A, Rudenko IN, Kaganovich A, Civiero L, Chau H, Kalia SK, Kalia LV, Lobbestael E, Chia R, Ndukwe K, Ding J, Nalls MA; International Parkinson's Disease Genomics Consortium; North American Brain Expression Consortium, Olszewski M, Hauser DN, Kumaran R, Lozano AM, Baekelandt V, Greene LE, Taymans JM, Greggio E, Cookson MR. (2014) Unbiased screen for interactors of leucine-rich repeat kinase 2 supports a common pathway for sporadic and familial Parkinson disease. Proc Natl Acad Sci U S A 111:2626
68. Piccoli G, Onofri F, Cirnaru MD, Kaiser CJ, Jagtap P, Kastenmüller A, Pischedda F, Marte A, von Zweydorf F, Vogt A, Giesert F, Pan L, Antonucci F, Kiel C, Zhang M, Weinkauf S, Sattler M, Sala C, Matteoli M, Ueffing M, Gloeckner CJ (2014) Leucine-rich repeat kinase 2 binds to neuronal vesicles through protein interactions mediated by its C-terminal WD40 domain. Mol Cell Biol 34:2147
69. Berwick DC, Harvey K (2011) LRRK2 signaling pathways: the key to unlocking neurodegeneration? Trends Cell Biol 21:257
70. Verma M, Steer EK, Chu CT (2013) ERKed by LRRK2: a cell biological perspective on hereditary and sporadic Parkinson's disease. Biochim Biophys Acta. doi:10.1016/j.bbadis.2013.11.005
71. Rideout HJ, Stefanis L (2014) The neurobiology of LRRK2 and its role in the pathogenesis of Parkinson's disease. Neurochem Res 39:576
72. Cookson MR (2010) The role of leucine-rich repeat kinase 2 (LRRK2) in Parkinson's disease. Nat Rev Neurosci 12:791
73. Hindle SJ, Elliott CJ (2013) Spread of neuronal degeneration in a dopaminergic, Lrrk-G2019S model of Parkinson disease. Autophagy 9:936
74. Yao C, Johnson WM, Gao Y, Wang W, Zhang J. Deak M, Alessi DR, Zhu X, Mieyal JJ, Roder H, Wilson-Delfosse AL, Chen SG (2013) Kinase inhibitors arrest neurodegeneration in cell and *C. elegans* models of LRRK2 toxicity. Hum. Mol Genet 22:328
75. Li X, Patel JC, Wang J, Avshalumov MV, Nicholson C, Buxbaum JD, Elder GA, Rice ME, Yue Z (2010) Enhanced striatal dopamine transmission and motor performance with LRRK2 overexpression in mice is eliminated by familial Parkinson's disease mutation G2019S. J Neurosci 30:1788
76. Li Y, Liu W, Oo TF, Wang L, Tang Y, Jackson-Lewis V, Zhou C, Geghman K, Bogdanov M, Przedborski S, Beal MF, Burke RE, Li C (2009) Mutant LRRK2(R1441G) BAC transgenic mice recapitulate cardinal features of Parkinson's disease. Nat Neurosci 12:826
77. Ramonet D, Daher JP, Lin BM, Stafa K, Kim J, Banerjee R, Westerlund M, Pletnikova O, Glauser L, Yang L, Liu Y, Swing DA, Beal MF, Troncoso JC, McCaffery JM, Jenkins NA, Copeland NG, Galter D, Thomas B, Lee MK, Dawson TM, Dawson VL, Moore DJ (2011) Dopaminergic neuronal loss, reduced neurite complexity and autophagic abnormalities in transgenic mice expressing G2019S mutant LRRK2. PLoS One 6:e18568
78. Lee BD, Shin JH, VanKampen J, Petrucelli L, West AB, Ko HS, Lee YI, Maguire-Zeiss KA, Bowers WJ, Federoff HJ, Dawson VL, Dawson TM (2010) Inhibitors of leucine-rich repeat kinase-2 protect against models of Parkinson's disease. Nat Med 16:998
79. Dusonchet J, Kochubey O, Stafa K, Young SM Jr, Zufferey R, Moore DJ, Schneider BL, Aebischer P (2011) A rat model of progressive nigral neurodegeneration induced by the Parkinson's disease-associated G2019S mutation in LRRK2. J Neurosci 31:07
80. Tieu K (2011) A guide to neurotoxic animal models of Parkinson's disease. Cold Spring Harb Perspect Med 1:a009316
81. Andres-Mateos E, Mejias R, Sasaki M, Li X, Lin BM, Biskup S, Zhang L, Banerjee R, Thomas B, Yang L, Liu G, Beal MF, Huso DL, Dawson TM, Dawson VL (2009) Unexpected lack of hypersensitivity in LRRK2 knock-out mice to MPTP (1-methyl-4-phenyl-1,2,3,6-tetrahydropyridine). J Neurosci 29:15846
82. Biskup S, Moore DJ, Rea A, Lorenz-Deperieux B, Coombes CE, Dawson VL, Dawson TM, West AB (2007) Dynamic and redundant regulation of LRRK2 and LRRK1 expression. BMC Neurosci 8:102

83. Tong Y, Yamaguchi H, Giaime E, Boyle S, Kopan R, Kelleher RJ, Shen J. (2010) Loss of leucine-rich repeat kinase 2 causes impairment of protein degradation pathways, accumulation of alpha-synuclein, and apoptotic cell death in aged mice. Proc Natl Acad Sci U S A. 107: 9879
84. Zhou H, Huang C, Tong J, Hong WC, Liu YJ, Xia XG (2011) Temporal expression of mutant LRRK2 in adult rats impairs dopamine reuptake. Int J Biol Sci 7:753
85. LEH-LRRK2tm1sage from SAGE Labs, product number: TGRL4620. http://www.sageresearchlabs.com/files/Parkinsons%20Disease%20KORat%20Flyer_1.pdf. Accessed 4 August 2014
86. Herzig MC, Kolly C, Persohn E, Theil D, Schweizer T, Hafner T, Stemmelen C, Troxler TJ, Schmid P, Danner S, Schnell CR, Mueller M, Kinzel B, Grevot A, Bolognani F, Stirn M, Kuhn RR, Kaupmann K, van der Putten PH, Rovelli G, Shimshek DR (2011) LRRK2 protein levels are determined by kinase function and are crucial for kidney and lung homeostasis in mice. Hum Mol Genet 20:4209
87. Ness D, Ren Z, Gardai S, Sharpnack D, Johnson VJ, Brennan RJ, Brigham EF, Olaharski AJ (2013) Leucine-rich repeat kinase 2 (LRRK2)-deficient rats exhibit renal tubule injury and perturbations in metabolic and immunological homeostasis. PLoS One 8:e66164
88. Miklavc P, Ehinger K, Thompson KE, Hobi N, Shimshek DR, Frick M (2010) Surfactant secretion in LRRK2 knock-out rats: changes in lamellar body morphology and rate of exocytosis. PLoS One 9:e84926
89. Baptista MA, Dave KD, Frasier MA, Sherer TB, Greeley M, Beck MJ, Varsho JS, Parker GA, Moore C, Churchill MJ, Meshul CK, Fiske BK (2013) Loss of leucine-rich repeat kinase 2 (LRRK2) in rats leads to progressive abnormal phenotypes in peripheral organs. PLoS One 8:e80705
90. Fuji, R (2013) Nonclinical safety studies of selective LRRK2 kinase inhibitors. Presented at Seventh annual Parkinson's disease therapeutics conference, The New York Academy of Sciences, October 24, 2013
91. The UniProt Consortium (2013) Update on activities at the Universal Protein Resource (UniProt) in 2013. Nucl Acids Res 41:D43
92. Bosgraaf L, Van Haastert PJM (2003) Roc, a Ras/GTPase domain in complex proteins. Biochim Biophys Acta 1643:5
93. Marin I, van Egmond WN, van Haastert PJM (2008) The Roco protein family: a functional perspective. Faseb J 22:3103
94. Marin I (2006) The Parkinson disease gene LRRK2: evolutionary and structural insights. Mol Biol Evol 23:2423
95. Mills RD, Mulhern TD, Cheng HC, Culvenor JG (2012) Analysis of LRRK2 accessory repeat domains: prediction of repeat length, number and sites of Parkinson's disease mutations. Biochem Soc Trans 40:1086
96. Tewari R, Bailes E, Bunting KA, Coates JC (2010) Armadillo-repeat protein functions: questions for little creatures. Trends Cell Biol 20:470
97. Mata IF, Wedemeyer WJ, Farrer MJ, Taylor JP, Gallo KA (2006) LRRK2 in Parkinson's disease: protein domains and functional insights. Trends Neurosci 29:286
98. Mosavi LK, Cammett TJ, Desrosiers DC, Zy P (2004) The ankyrin repeat as molecular architecture for protein recognition. Protein Sci 13:1435
99. Li J, Mahajan A, Tsai MD (2006) Ankyrin repeat: a unique motif mediating protein–protein interactions. Biochemistry 45:15168
100. Vancraenenbroeck R, Lobbestael E, Weeks SD, Strelkov SV, Baekelandt V, Taymans JM, DeMaeyer M (2012) Expression, purification and preliminary biochemical and structural characterization of the leucine rich repeat namesake domain of leucine rich repeat kinase 2. Biochim Biophys Acta 1824:450
101. Guo L, Wang W, Chen SG (2006) Leucine-rich repeat kinase 2: relevance to Parkinson's disease. Int J Biochem Cell Biol 38:1469
102. Kobe B, Kajava AV (2001) The leucine-rich repeat as a protein recognition motif. Curr Opin Struct Biol 11:725

103. Bella J, Hindle KL, McEwan PA, Lovell SC (2008) The leucine-rich repeat structure. Cell Mol Life Sci 65:2307
104. Xu C, Min J (2011) Structure and function of WD40 domain proteins. Protein Cell 2:202
105. Taymans JM (2012) The GTPase function of LRRK2. Biochem Soc Trans 40:1063
106. Xiong Y, Dawson VL, Dawson TM (2012) LRRK2 GTPase dysfunction in the pathogenesis of Parkinson's disease. Biochem Soc Trans 40:1074
107. Tsika E, Moore D (2013) Contribution of GTPase activity to LRRK2-associated Parkinson disease. Small GTPases 4:164
108. Deng J, Lewis PA, Greggio E, Sluch E, Beilina A, Cookson MR (2008) Structure of the ROC domain from the Parkinson's disease-associated leucine-rich repeat kinase 2 reveals a dimeric GTPase. Proc Natl Acad Sci U S A 105:1499
109. Gotthardt K, Weyand M, Kortholt A, VanHaastert PJM, Wittinghofer A (2008) Structure of the Roc-COR domain tandem of C. tepidum, a prokaryotic homologue of the human LRRK2 Parkinson kinase. EMBO J 27:2239
110. Gotthardt K, Weyand M, Kortholt A, VanHaastert PJ, Wittinghofer A (2008) Structure of the Roc-COR domain tandem of C. tepidum, a prokaryotic homologue of the human LRRK2 Parkinson kinase. EMBO J 27:2352
111. Gasper R, Meyer S. Gotthardt K, Sirajuddin M, Wittinghofer A (2009) It takes two to tango: regulation of G proteins by dimerization. Nat Rev Mol Cell Biol 10:423
112. Greggio E, Cookson MR (2009) Leucine-rich repeat kinase 2 mutations and Parkinson's disease: three questions. ASN Neuro 1:13
113. Li Y, Dunn L, Greggi E, Krumm B, Jackson GS, Cookson MR, Lewis MR, Deng J (2009) The R1441C mutation alters the folding properties of the ROC domain of LRRK2. Biochim Biophys Acta 1792:1194
114. Ray S, Liu M (2012) Current understanding of LRRK2 in Parkinson's disease: biochemical and structural features and inhibitor design. Future Med Chem 4:1701
115. Manning G, Whyte DB, Martinez R, Hunter T, Sudarsanam S (2002) The protein kinase complement of the human genome. Science 298:1912
116. Zhang D, Lin J, Han J (2010) Receptor-interacting protein (RIP) kinase family. Cell Mol Immunol 7:243
117. Adams JA (2001) Kinetic and catalytic mechanisms of protein kinases. Chem Rev 101:2271
118. Gilsbach BK, Ho FY, Vetter IR, vanHaastert PJM, Wittinghofer A, Kortholt A (2012) Roco kinase structures give insights into the mechanism of Parkinson disease-related leucine-rich-repeat kinase 2 mutations. Proc Natl Acad Sci U S A 109:10322
119. Aasly JO, Toft M, Fernandez-Mata I, Kacherqus J, Hulihan M, White LR, Farrer M (2005) Clinical features of LRRK2-associated Parkinson's disease in central Norway. Ann Neurol 57:762
120. Bardien S, Lesage S, Brice A, Carr J (2011) Genetic characteristics of leucine-rich repeat kinase 2 (LRRK2) associated Parkinson's disease. Parkinsonism Relat Disord 17:501
121. Paisan-Ruiz C (2009) LRRK2 gene variation and its contribution to Parkinson disease. Hum Mutat 30:1153
122. Paisan-Ruiz C, Lewis PA, Singleton AB (2013) LRRK2: cause, risk, and mechanism. J Parkinson's Dis 3:85
123. Cavasotto CN, Phatak SS (2009) Homology modeling in drug discovery: current trends and applications. Drug Discov Today 14:676
124. Krieger E, Nabuurs SB, Vriend G (2005) Homology modeling. In: Bourne PE, Weissig H (eds) Structural bioinformatics, vol 44. Wiley, Hoboken. doi:10.1002/0471721204.ch25
125. LRRK2 homology models were constructed with MOE (Molecular Operating Environment (MOE), 2013.08; Chemical Computing Group Inc., 1010 Sherbooke St. West, Suite #910, Montreal, QC, Canada, H3A 2R7, 2013) using the AMBER99 force field (Wang JM, Cieplak P, Kollman PA. (2000) How well does a restrained electrostatic potential (RESP) model perform in calculating conformational energies of organic and biological molecules? J Comput Chem 21:1049). The amino acid sequence of human LRRK2 was retrieved from

UniProt (The UniProt Consortium (2008) The Universal Protein Resource (UniProt) Nucleic Acids Res 36:D190) and aligned to the template structure sequences specified in the text using MOE's Protein Align function with default settings followed by manual editing of loop regions, insertions and deletions
126. Daniels V, Vancraenenbroeck R, Law BMH, Greggio E, Lobbestael E, Gao F, DeMaeyer M, Cookson MR, Harvey K, Baekelandt V, Taymans JM (2011) Insight into the mode of action of the LRRK2 Y1699C pathogenic mutant. J Neurochem 116:304
127. Fiegen D, Dvorsky R, Ahmadian M (2006) Structural principles of Ras interaction with regulators and effectors. In: Der C (ed) RAS family GTPases, Springer, The Netherlands, p 45
128. Rojas J, Ras-Gefs E, Santos (2006) Ras Gaps. In: Der C (ed) RAS family GTPases, Springer, the Netherlands, p 15
129. Albrecht M (2005) LRRK2 mutations and Parkinsonism. Lancet 365:1230
130. Liu M, Kang S, Ray S, Jackson J, Zaitsev AD, Gerber SA, Cuny GD, Glicksman MA (2011) Kinetic, mechanistic, and structural modeling studies of truncated wild-type leucine-rich repeat kinase 2 and the G2019S mutant. Biochemistry 50:9399
131. Liu M, Bender SA, Cuny GD, Sherman W, Glicksman M, Ray SS (2013) Type II kinase inhibitors show an unexpected inhibition mode against Parkinson's disease-linked LRRK2 mutant G2019S. Biochemistry 52:1725
132. Chen H, Chan BK, Drummond J, Estrada AA, Gunzer-Toste J, Liu X, Liu Y, Moffat J, Shore D, Sweeney ZK, Tran T, Wang S, Zhao G, Zhu H, Burdick DJ (2012) Discovery of selective LRRK2 inhibitors guided by computational analysis and molecular modeling. J Med Chem 55:5536
133. Estrada AA, Liu X, Baker-Glenn C, Beresford A, Burdick DJ, Chambers M, Chan BK, Chen H, Ding X, DiPasquale AG, Dominguez SL, Dotson J, Drummond J, Flagella M, Flynn S, Fuji R, Gill A, Gunzner-Toste J, Harris SF, Heffron TP, Kleinheinz T, Lee DW, LePichon DE, Lyssikatos JP, Medhurst AD, Moffat JG, Mukund S, Nash K, Scearce-Levie K, Shang Z, Shore DG, Tran T, Tivedi N, Wang S, Zhang S, Zhang X, Zhao G, Zhu H, Sweeney ZK (2012) Discovery of highly potent, selective, and brain-penetrable leucine-rich repeat kinase 2 (LRRK2) small molecule inhibitors. J Med Chem 55:9416
134. Chan BK, Estrada AA, Chen H, Atherall J, Baker-Glenn C, Beresford A, Burdick DJ, Chambers M, Dominguez SL, Drummond J, Gill A, Keinheinz T, LePichon CE, Medhurst AD, Liu X, Moffat JG, Nash K, Scearce-Levie K, Sheng Z, Shore DG, VandePoel H, Zhang S, Zhu H, Sweeney ZK (2013) Discovery of a highly selective, brain-penetrant aminopyrazole LRRK2 inhibitor. ACS Med Chem Lett 4:85
135. Estrada AA, Chan BK, Baker-Glenn C, Beresford A, Burdick DJ, Chambers M, Chen H, Dominguez SL, Dotson J, Drummond J, Flagella M, Fuji R, Gill A, Halladay J, Harris SF, Heffron TP, Kleinheinz T, Lee DW, LePichon CR, Liu X, Lyssikatos JP, Medhurst AD, Moffat JG, Nash K, Scearce-Levie K, Sheng Z, Shore DG, Wong S, Zhang S, Zhang X, Zhu H, Sweeney ZK (2014) Discovery of highly potent, selective, and brain-penetrant aminopyrazole leucine-rich repeat kinase 2 (LRRK2) small molecule inhibitors. J Med Chem 57:921
136. Drolet RE, Sanders JM, Kern JT (2011) Leucine-rich repeat kinase 2 (LRRK2) cellular biology: a review of recent advances in identifying physiological substrates and cellular functions. J Neurogenet 25:140
137. Franzini M, Ye ZM, Adler M, Aubele DL, Garofalo AW, Gauby S, Goldbach E, Probst GD, Quinn KP, Santiago P, Sham HL, Tam D, Troung A, Ren Z (2013) Triazolopyridazine LRRK2 kinase inhibitors. Bioorg Med Chem Lett 23:1967
138. Garofalo AW, Adler M, Aubele DL, Brigham EF, Chian D, Franzini M, Goldbach E, Kwong GT, Motter R, Probst GD, Quinn KP, Ruslim L, Sham HL, Tam D, Tanaka P, Troung AP, Ye XM, Ren Z (2013) Discovery of 4-alkylamino-7-aryl-3-cyanoquinoline LRRK2 kinase inhibitors. Bioorg Med Chem Lett 23:1974

139. Garofalo AW, Adler M, Aubele DL, Bowers S, Franzini M, Goldbach E, Lorentzen C, Neitz RJ, Probst GD, Quinn KP, Santiago P, Sham HL, Tam D, Troung AP, Ye XM, Ren Z (2013) Novel cinnoline-based inhibitors of LRRK2 kinase activity. Bioorg Med Chem Lett 23:71
140. Deng X, Elkins JM, Zhang J, Yang Q, Erazo T, Gomez N, Choi HG, Wang J, Dzamko N, Lee JD, Sim T, Kim N, Alessi DR, Lizcano JM, Knapp S, Gray NS (2013) Structural determinants for ERK5 (MAPK7) and leucine rich repeat kinase 2 activities of benzo[e]pyrimido-[5,4-b] diazepine-6(11H)-ones. Eur J Med Chem 70C:758
141. Anand VS, Braithwaite SP (2009) LRRK2 in Parkinson's disease: biochemical functions. Febs J 276:6428
142. Yun H, Heo HY, Kim HH, DooKim N, Seol W (2011) Identification of chemicals to inhibit the kinase activity of leucine-rich repeat kinase 2 (LRRK2), a Parkinson's disease-associated protein. Bioorg Med Chem Lett 21:2953
143. Deng X, Dzamko N, Prescott A, Davies P, Liu Q, Yang Q, Lee JD, Patricelli MP, Nomanbhoy TN, Alessi DR, Gray NS (2011) Characterization of a selective inhibitor of the Parkinson's disease kinase LRRK2. Nat Chem Biol 7:203
144. Troxler T, Greenidge P, Zimmerman K, Desrayaud S, Drukes P, Schweizer T, Stauffer D, Rovelli G, Shimshek DR (2013) Discovery of novel indolinone-based, potent, selective and brain penetrant inhibitors of LRRK2. Bioorg Med Chem Lett 23:4085
145. Choi HG, Zhang J, Deng X, Hatcher JM, Patricelli MP, Zhao Z, Aless DR, Gray NS (2012) Brain penetrant LRRK2 inhibitor. ACS Med Chem Lett 3:658
146. Zhang J, Deng X, Choi HG, Alessi DR, Gray NS (2012) Characterization of TAE684 as a potent LRRK2 kinase inhibitor. Bioorg Med Chem Lett 22:1864
147. Reith AD, Bamborough P, Jandu K, Andreotti D, Mensah L, Dossang P, Choi HG, Deng X, Zhang J, Alessi DR, Gray NS (2012) GSK2578215A; a potent and highly selective 2-arylmethyloxy-5-substitutent-N-arylbenzamide LRRK2 kinase inhibitor. Bioorg Med Chem Lett 22:5625
148. Liu M, Poulose S, Schuman E, Zaitsev AD, Dobson B, Auerbach K, Seyb K, Cuny GD, Glicksman MA, Stein RL, Yue Z (2010) Development of a mechanism-based high-throughput screen assay for leucine-rich repeat kinase 2-Discovery of LRRK2 inhibitors. Anal Biochem 404:186
149. Wager TT, Hou X, Verhoest PR, Villalobos A (2010) Moving beyond rules: the development of a central nervous system multiparameter optimization (CNS MPO) approach to enable alignment of druglike properties. ACS Chem Neurosci 1:435
150. Labbe C, Ross OA (2014) Associated studies of sporadic Parkinson's disease in the genomic era. Curr Genomics 15:2
151. Hughes JD, Blagg J, Price DA, Bailey S, Decrscenzo GA, Devraj RV, Ellsworth E, Fobian YM, Gibbs ME, Gilles RW, Greene N, Huang E, Krieger-Burke T, Loesel J, Wager T, Zhang Y (2008) Physiochemical drug properties associated with in vivo toxicological outcomes. Bioorg Med Chem Lett 18:4872
152. Edwards AM, Isserlin R, Bader GD, Frye SV, Willson TM, Yu FH (2011) Too many roads not taken. Nature 470:163
153. Liou GY, Gallo KA (2009) New biochemical approaches towards understanding Parkinson's disease-associated kinase, LRRK2 Biochem. J 424:e1
154. Covy JP, Giasson BI (2009) Identification of compounds that inhibit the kinase activity of leucine-rich repeat kinase 2. Biochem Biophys Res Commun 378:473
155. Davis MI, Hunt JP, Herrgard S, Ciceri P, Wodicka LM, Pallares G, Hocker M, Treiber DK, Zarrinkar PP (2011) Comprehensive analysis of kinase inhibitor selectivity. Nat Biotech 29:1046
156. Ramsden N, Perrin J, Ren Z, Lee BD, Zinn N, Dawson VL, Tam D, Bova M, Lang M, Drewes G, Bantscheff M, Bard F, Dawson TM, Hopf C (2011) Chemoproteomics-based design of potent LRRK2-selective lead compounds that attenuate Parkinson's disease-related toxicity in human neurons. ACS Chem Biol 6:1021

157. Lee J, Song HJ, Koh JS, Lee HK, Kim Y, Chang S, Kim HW, Lim SH, Choi JS, Lim SH, Kim SW (2011) Patent Application WO2011060295
158. Baker-Glenn C, Burdick DJ, Chambers M, Chan BK, Chen H, Estrada, A, Guzner JL, Shore D, Sweeney ZK, Wang S, Zhao G (2011) Patent Application WO2011151360
159. Nicols PL, Eatherton AJ, Bamborough P, Jandu KJ, Phillips OJ, Andreotti D (2011) Patent Application WO2011038572
160. Galatsis P, Hayward MM, Kormos BL, Wager TT, Zhang L, Stepan AF, Henderson JL, Kurumbail RG, Verhoest PR (2014) Patent Application WO2014001973
161. Di L, Whitney-Pickett C, Umland JP, Zhang H, Zhang X, Gebhard DF, Lai Y, Federico JJ, Davidson RE, Smith R, Reyner EL, Lee C, Feng B, Rotter C, Varma MV, Kempshall S, Fenner K, El-kattan AF, Liston TE, Troutman MD (2011) Development of a new permeability assay using low-efflux MDCKII cells. J Pharm Sci 100:4974
162. Feng B, Mills JB, Davidson RE, Mireles RJ, Janiszewski JS, Troutman MD, de Morais SM (2008) In vitro P-glycoprotein assays to predict the in vivo interactions of P-glycoprotein with drugs in the central nervous system. Drug Metab Disposit 36:268
163. Finlayson K, Sharkey J (2004) A High-throughput binding assay for HERG. In: Yan Z, Caldwell GW (eds) Methods in pharmacology and toxicology. Humana, Totowa, pp 353–368
164. Greene N, Aleo MD, Louise-May S, Price DA, Will Y (2010) Using an in vitro cytotoxicity assay to aid in compound selection for in vivo safety studies. Bioorg Med Chem Lett 20:5308
165. Marroquin LD, Hynes J, Dykens JA, Jamieson JD, Will Y (2007) Circumventing the Crabtree effect: replacing media glucose with galactose increases susceptibility of HepG2 cells to mitochondrial toxicants. Toxicol Sci 97:539
166. Dundee Panel Data (2014) http://www.kinase-screen.mrc.ac.uk/kinase-inhibitors. Accessed 2 June 2014
167. Okerberg ES, Wu J, Zhang B, Samii B, Blackford K, Winn DT, Shreder KR, Burbaum JJ, Patricelli MP (2005) High-resolution functional proteomics by active-site peptide profiling. Proc Natl Acad Sci 102:4996

Index